Springer Tracts in Modern Physics
Volume 150

Springer-Verlag Berlin Heidelberg GmbH

Springer Tracts in Modern Physics

Springer Tracts in Modern Physics provides comprehensive and critical reviews of topics of current interest in physics. The following fields are emphasized: elementary particle physics, solid-state physics, complex systems, and fundamental astrophysics.

Suitable reviews of other fields can also be accepted. The editors encourage prospective authors to correspond with them in advance of submitting an article. For reviews of topics belonging to the above mentioned fields, they should address the responsible editor, otherwise the managing editor.
See also http://www.springer.de/phys/books/stmp.html

Managing Editor

Gerhard Höhler

Institut für Theoretische Teilchenphysik
Universität Karlsruhe
Postfach 69 80
D-76128 Karlsruhe, Germany
Phone: +49 (7 21) 6 08 33 75
Fax: +49 (7 21) 37 07 26
Email: gerhard.hoehler@physik.uni-karlsruhe.de
http://www-ttp.physik.uni-karlsruhe.de/

Elementary Particle Physics, Editors

Johann H. Kühn

Institut für Theoretische Teilchenphysik
Universität Karlsruhe
Postfach 69 80
D-76128 Karlsruhe, Germany
Phone: +49 (7 21) 6 08 33 72
Fax: +49 (7 21) 37 07 26
Email: johann.kuehn@physik.uni-karlsruhe.de
http://www-ttp.physik.uni-karlsruhe.de/~jk

Thomas Müller

Institut für Experimentelle Kernphysik
Fakultät für Physik
Universität Karlsruhe
Postfach 69 80
D-76128 Karlsruhe, Germany
Phone: +49 (7 21) 6 08 35 24
Fax: +49 (7 21) 6 07 26 21
Email: thomas.muller@physik.uni-karlsruhe.de
http://www-ekp.physik.uni-karlsruhe.de

Roberto Peccei

Department of Physics
University of California, Los Angeles
405 Hilgard Avenue
Los Angeles, CA 90024-1547, USA
Phone: +1 310 825 1042
Fax: +1 310 825 9368
Email: peccei@physics.ucla.edu
http://www.physics.ucla.edu/faculty/ladder/
peccei.html

Solid-State Physics, Editor

Peter Wölfle

Institut für Theorie der Kondensierten Materie
Universität Karlsruhe
Postfach 69 80
D-76128 Karlsruhe, Germany
Phone: +49 (7 21) 6 08 35 90
Fax: +49 (7 21) 69 81 50
Email: woelfle@tkm.physik.uni-karlsruhe.de
http://www-tkm.physik.uni-karlsruhe.de

Complex Systems, Editor

Frank Steiner

Abteilung Theoretische Physik
Universität Ulm
Albert-Einstein-Allee 11
D-89069 Ulm, Germany
Phone: +49 (7 31) 5 02 29 10
Fax: +49 (7 31) 5 02 29 24
Email: steiner@physik.uni-ulm.de
http://www.physik.uni-ulm.de/theo/theophys.html

Fundamental Astrophysics, Editor

Joachim Trümper

Max-Planck-Institut für Extraterrestrische Physik
Postfach 16 03
D-85740 Garching, Germany
Phone: +49 (89) 32 99 35 59
Fax: +49 (89) 32 99 35 69
Email: jtrumper@mpe-garching.mpg.de
http://www.mpe-garching.mpg.de/index.html

Michael Kuhlen

QCD at HERA

The Hadronic Final State in Deep Inelastic Scattering

With 99 Figures

 Springer

Dr. Michael Kuhlen
Max-Planck Institut für Physik
Werner-Heisenberg-Institut
Föhringer Ring 6
D-80805 München
Email: michael.kuhlen@desy.de

Physics and Astronomy Classification Scheme (PACS):
12.38.-t, 13.60.-r, 13.87.-a

ISSN 0081-3869
ISBN 978-3-662-14733-7

Library of Congress Cataloging-in-Publication Data applied for. Die Deutsche Bibliothek – CIP Einheitsaufnahme

Kuhlen, Michael: QCD at HERA: the hadronic final state in deep inelastic scattering / Michael Kuhlen.

(Springer tracts in modern physics; Vol. 150)

ISBN 978-3-662-14733-7   ISBN 978-3-540-49641-0 (eBook)
DOI 10.1007/978-3-540-49641-0

The use of general descriptive names, registered names, trademarks, etc. in this publication does not imply, even in the absence of a specific statement, that such names are exempt from the relevant protective laws and regulations and therefore free for general use.

Typesetting: Camera-ready copy by the author using a Springer TeX macro package
Cover design: *design & production* GmbH, Heidelberg

SPIN: 10695920   56/3144 - 5 4 3 2 1 0 – Printed on acid-free paper

Preface

HERA[1] is the world's first electron–proton collider.[2] The large centre-of-mass (CM) energy of 300 GeV allows the exploration of new regimes: new particles with masses up to 300 GeV can be produced, the structure of the proton can be studied with a resolving power varying over 5 orders of magnitude down to dimensions of 10^{-18} m, and partons with very small fractional proton momenta (Bjorken x down to 10^{-6}) become experimentally accessible. The hadronic final state which emerges when the proton breaks up has an invariant mass of up to 300 GeV. It provides a laboratory to study quantum chromodynamics (QCD) under varying experimental conditions, which can be controlled by measuring the scattered electron. The hadronic final state carries information on the structure of the proton. It is made use of to study the dynamics of the proton's constituents, complementary to structure function measurements.

In contrast to "clean" e^+e^- interactions, the initial state in ep collisions already contains a strongly interacting particle. That makes the physics more complicated, but leads also to interesting effects that cannot be studied in e^+e^- collisions. At HERA, well-established fields are being studied, exploiting the tunable kinematic conditions. These comprise jet physics and comparisons with perturbative QCD to extract the strong coupling α_s and the density of gluons in the proton, or for example the measurement of fragmentation functions. Other topics have only blossomed with the advent of HERA. To name a few, a class of events with large rapidity gaps and small x physics allow longstanding problems in QCD to be addressed, which are connected with scattering cross-sections at high energies and confinement. There are also exotic effects like instanton-induced reactions, which, if discovered at HERA, would alter significantly our view of particle physics.

Purpose

This work provides a review of the hadronic final state measurements at HERA in deep inelastic scattering (DIS). The emphasis is on experimental results, because in many cases at HERA, experiment is driving theory.

---

[1] "Hadron–Elektron–Ring–Anlage"

[2] HERA can operate with either electrons or positrons. In the following, the generic name electron is used for electrons as well as for positrons.

Many measurements are being performed without any theoretical prediction, and often they have not yet found an unambiguous theoretical explanation. Nevertheless, the results are discussed in the context of the theory, where possible.

The review of the experimental situation is complete up to the fall of 1997, with a few updates concerning the situation in summer 1998. It can therefore be consulted for quick access to the HERA data. Basic concepts are explained in a partly pedagogical fashion to serve physicists from outside the HERA community and newcomers to HERA physics.

Contents

In Chap. 1 the HERA machine and the H1 and ZEUS experiments are introduced, and the kinematic variables are discussed. In Chap. 2 the theoretical framework of deep inelastic scattering is set up with evolution equations and a special section on the interest in small x physics. The hadronic final state should not be discussed without knowledge of the inclusive ep cross-section and the proton structure functions (Chap. 3). In Chap. 4 we start with simple models for hadron production. They are subsequently refined in and serve as the basis for the discussion of the data.

Measurements of basic event properties are presented in Chap. 5: energy flows, charged particle spectra, charm and strangeness contents, and Bose–Einstein correlations. The fragmentation of the scattered quark is studied in Chap. 6 and compared with quark fragmentation in e^+e^- annihilation and QCD calculations. Measurements of event shape variables offer a new view of hadronization properties with "power corrections", offering a potentially powerful tool for measurements of the strong coupling α_s. Jet production has been compared to perturbative QCD predictions to measure α_s and the gluon density in the proton (Chap. 7). By measuring jet rates, regions of phase space have been identified where the measured jet rates are not well understood yet, and where the underlying physics possibly departs from the conventional picture of deep inelastic scattering.

Also the energy flow measurements at small x have not yet found an unambiguous theoretical interpretation. In Chap. 8 on low x physics dedicated searches for "footprints" of new QCD effects (BFKL) are discussed: energy flows, high p_T particles and "forward jets". Chapter 9 deals with the possibility of discovering QCD instantons at HERA, which would have far reaching consequences for our understanding of field theories and for cosmology.

Some related topics and neighbouring fields could only be touched upon. The interested reader is referred to other reviews on structure functions [1, 2], rapidity gaps [3], photoproduction [4] and on hadron production at fixed target experiments [5].

Throughout, unless stated otherwise, the data shown have been corrected for detector effects and QED radiation, and the errors comprise statistical and systematic errors added in quadrature.

Acknowledgements

This work has been made possible by a grant from the Deutsche Forschungsgemeinschaft, which allowed me to write a "Habilitation". The result is this book. I would like to thank G. Buschhorn from the Max-Planck-Institut für Physik in Munich for the hospitality I have experienced in his group during the last two years. I have enjoyed very much the close collaboration in the MPI group, in particular with F. Botterweck, T. Carli and G. Grindhammer.

I wish to thank all my colleagues from the H1 and ZEUS experiments, on whose research this work is based, for their collaboration and friendly competition at HERA. I would like to thank J. Dainton, E. Lohrmann, B. Naroska and A. Wagner for their continuous encouragement and their support to undertake the Habilitation and to teach at the Universität Hamburg.

I am grateful to J. Bartels, E. De Wolf, A. Martin, G. Ingelman, A. Ringwald and F. Schrempp for their patient explanations of sometimes heavy theory, and for enlightening discussions, in which some very fruitful ideas were born. Helpful discussions with B. Andersson, Y. Dokshitzer, D. Graudenz, L. Lönnblad, M. Seymour, S. Lang, E. Mirkes and B. Webber are gratefully acknowledged.

Finally, I would like to thank J. Bartels, N. Brook, T. Carli, T. Doyle, A. De Roeck, E. De Wolf, G. Grindhammer, J. Hartmann, G. Ingelman, H. Jung, A. Levy, E. Lohrmann, U. Martyn, D. Milstead, B. Naroska, N. Pavel, J. Repond, A. Ringwald, F. Schrempp, F. Sefkow and V. Shekelyan for their comments on the manuscript, or parts of it. Their critical reading has been invaluable.

Hamburg, August 1998   *Michael Kuhlen*

Contents

1. Introduction

1.1 The HERA Machine

In the HERA machine, electrons and and protons are accelerated and stored in two separate rings. The circumference of the machine is 6.3 km. The magnets of the proton ring are superconducting, the magnets of the electron ring are conventional. The final beam energies are $E_e = 27.5$ GeV for electrons and $E_p = 820$ GeV for protons with a collision centre-of-mass energy of $\sqrt{s} = \sqrt{4E_e E_p} = 300$ GeV. Early data were collected with $E_e = 26.7$ GeV.

The beams are collided head-on in two interaction regions occupied by the H1 and ZEUS experiments. There are 220 bunch positions in the beam, of which typically 190 are filled with a few times 10^{10} particles per bunch. The time between bunch crossings is 96 ns. The longitudinal bunch length is about 60 cm, leading to an approximately Gaussian distribution of interaction points along the beam line with width 10 cm. The transverse beam size is 300 μm horizontally by 70 μm vertically.

In 1997 the average peak luminosity was $8.4 \cdot 10^{30}$ cm^{-2} s^{-1} with average beam currents at the beginning of a fill of 77 mA for protons and 36 mA for electrons. The total integrated luminosity for the 1997 run was 35 pb^{-1}. Most analyses in this review are based on data from the 1992–1994 runs, corresponding to an integrated luminosity of $\mathcal{O}(1-2)$ pb^{-1}.

1.2 The H1 and ZEUS Detectors

The H1 [6, 7] and ZEUS detectors [8] serve to detect the scattered electron in ep collisions and to measure the emerging hadrons. The individual detector components are mounted concentrically around the beam line. Due to the asymmetric beam energies, the hadronic system is boosted into the proton direction $(+z)$. Therefore the detectors are also asymmetric with respect to the interaction point, with enhanced instrumentation for hadrons in the $+z$ (forward) direction. The acceptances and resolutions of the main detector components for the analyses presented here are given in Table 1.1. Figure 1.1 shows a drawing of the ZEUS detector, and Fig. 1.2 an event display from H1.

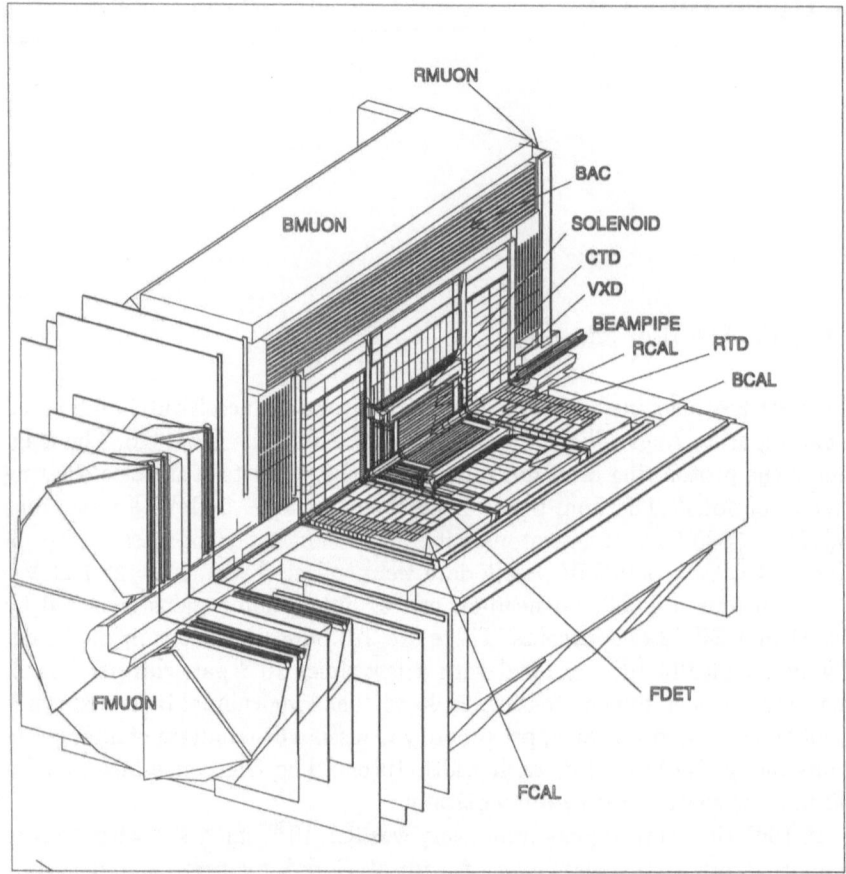

Fig. 1.1. The ZEUS detector. Protons come towards the observer. Shown are central, forward, backward and vertex tracking detectors (CTD, FDET, RTD, VXD), the uranium calorimeter (BCAL, RCAL, FCAL), the muon system (RMUON, BMUON, FMUON) and the backing calorimeter (BAC). The dimensions of the whole detector are roughly $10 \times 10 \times 18$ m^3

Closest to the beam line are wire chambers for measuring charged particle trajectories. The particles' momenta are determined from their track curvature in a longitudinal magnetic field provided by a superconducting coil.

Electromagnetic and hadronic showers are measured in calorimeters surrounding the inner tracking devices. H1 emphasizes electron detection with a finely segmented lead (inner electromagnetic part) and steel (outer hadronic part) liquid argon calorimeter (LAr) with good energy resolution for electrons, supplemented by a dedicated electromagnetic backward calorimeter. From 1992 to 1994 an electromagnetic lead/scintillator sandwich calorimeter was installed in the backward region (BEMC), and from 1995 onwards a lead/scintillating fibre calorimeter (SPACAL). A copper calorimeter with

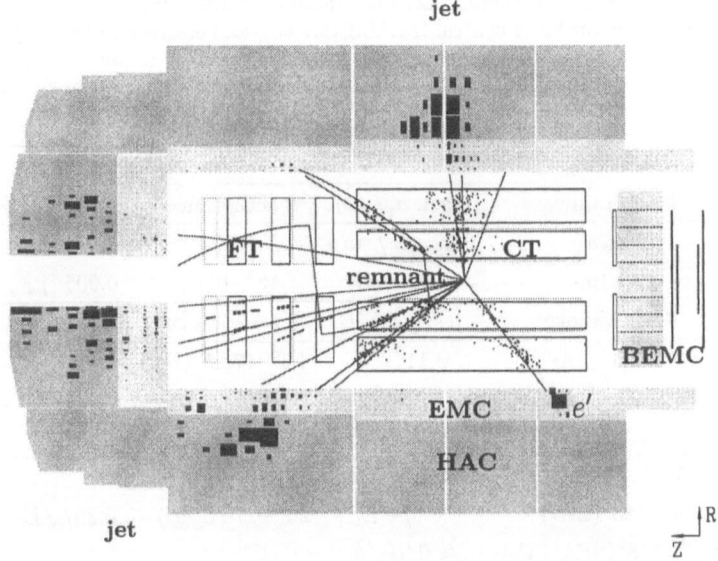

Fig. 1.2. DIS event with the scattered electron e' and two well separated jets detected in the H1 detector. The proton remnant leaves mostly undetected in the $+z$ direction. Shown are the central and forward tracking chambers (CT,FT), the electromagnetic and hadronic sections of the liquid argon calorimeter (EMC, HAC), and the backward electromagnetic calorimeter (BEMC)

silicon readout (PLUG) covers part of the forward beam hole. Compensation for the different response to hadronic and electromagnetic showers in the LAr is done offline by a software weighting technique. ZEUS achieves better hadronic energy resolution with a self-compensating uranium/scintillator calorimeter (U), and compromises on electromagnetic energy resolution.

The calorimeters are surrounded by chambers and absorber plates for measuring shower leakage and for muon detection. Further specialized detectors are installed very close to the beam line to detect particles that are scattered under small angles. Silicon detectors that have already been installed as vertex detectors or are being planned have not yet been used in physics analyses.

1.3 Kinematics

Definition of Kinematic Variables

The kinematics of the basic ep scattering process in Fig. 1.3 can be characterized by any set of two Lorentz invariants out of Q^2, x, y and W, which are built from the 4-momentum transfer $q = k - k'$ mediated by the virtual boson

Table 1.1. Acceptances and resolutions of the main detector components from H1 and ZEUS, namely forward and central tracking devices, and electromagnetic (e.m.) and hadronic calorimeters. In the resolution formulae, energy E and transverse momentum p_T are to be taken in GeV. Additional constant contributions to the resolutions of $\mathcal{O}(1-3\%)$ are not shown

	H1		ZEUS	
tracking	θ acceptance	resol. σ_{p_T}/p_T	θ acceptance	resol. σ_{p_T}/p_T
forward	7°–25°	$0.02 \cdot p_T/\sin\theta$	7.5°–28°	
central	20°–160°	$0.009 \cdot p_T$	15°–164°	$0.005 \cdot p_T$
calorim.	θ acceptance	resol. σ_E/E	θ acceptance	resol. σ_E/E
e.m.	4°–153°(LAr)	$0.11/\sqrt{E}$	2.2°–176.5°(U)	$0.18/\sqrt{E}$
	155.5°–174.5°(BEMC)	$0.1/\sqrt{E}$		
	151°–177.5°(SPACAL)	$0.075/\sqrt{E}$		
hadr.	0.7°–3.3°(PLUG)	$\approx 1.5/\sqrt{E}$		
	4°–153°(LAr)	$0.5/\sqrt{E}$	2.2°–176.5°(U)	$0.35/\sqrt{E}$
	153°–178°(SPACAL)	$\approx 0.3/\sqrt{E}$		

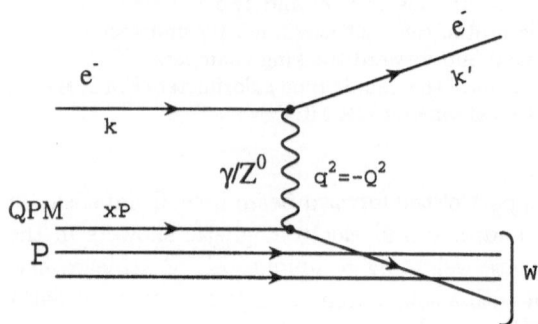

Fig. 1.3. Basic diagram for DIS in $O(\alpha_s^0)$ (quark parton model, QPM)

and from the 4-momentum P of the incoming proton. The ep invariant mass squared is $s = (e + P)^2$. These kinematic variables are then:

$$Q^2 := -q^2, \tag{1.1}$$

which gives the transverse resolving power of the probe with wavelength $\lambda = 1/Q$ (we set $\hbar = c = 1$);

$$x := \frac{Q^2}{2Pq}, \tag{1.2}$$

the Bjorken scaling variable ($0 \leq x \leq 1$), which can be interpreted as the momentum fraction of the proton which is carried by the struck quark (in a frame where the proton is fast, and assuming the quark–parton model to be a good approximation);

$$y := \frac{Pq}{Pk} = \frac{Q^2}{x(s - m_p^2)}, \tag{1.3}$$

the transferred energy fraction from the electron to the proton in the proton rest frame $(0 \leq y \leq 1)$; and

$$W^2 = Q^2 \frac{1 - x}{x} + m_p^2 \approx sy - Q^2, \tag{1.4}$$

the invariant mass squared of the outgoing hadronic system H. The invariant

$$\nu := \frac{Pq}{m_p}, \tag{1.5}$$

is rarely used at HERA. In the proton rest frame it gives the energy transfer from the lepton to the proton.

The large CM energy gives access to kinematic regions both at very small x and at large Q^2 (Fig. 1.4). The HERA data cover roughly $Q^2 = 0.2$ to 10^4 GeV^2, $x = 10^{-5}$ to 10^{-1} and $W = 40$ to 300 GeV.

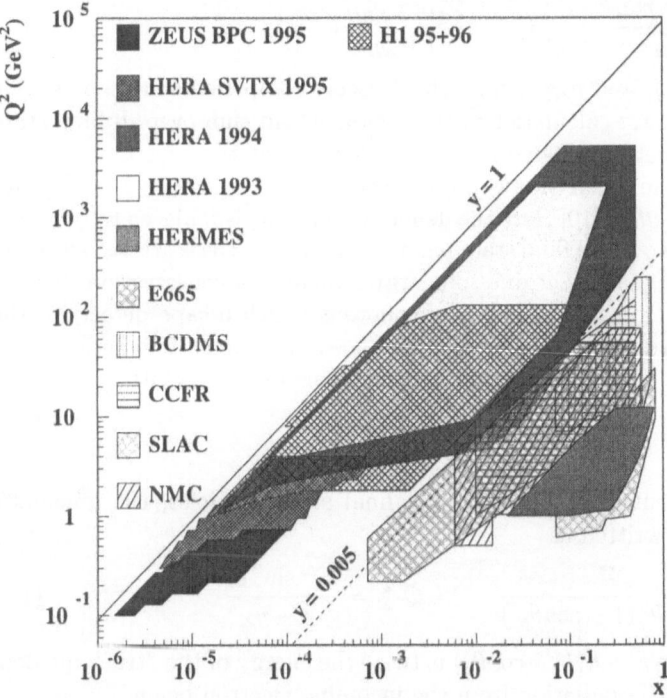

Fig. 1.4. Coverage in the kinematic plane (x, Q^2) of various DIS experiments. The kinematic boundary for HERA is given by the line $y = 1$. There exist also HERA data in the region 10^4 GeV$^2 < Q^2 < 10^5$ GeV^2

Experimental Reconstruction of the Event Kinematics

The kinematics can be determined either from the electron alone, or from the measured hadronic system alone, or from a combination of both, permitting important systematic cross-checks. The hadronic measurement relies mostly upon the calorimeters. At large y the precision of the hadron method can be improved by momentum measurements in the trackers.

The electron method. The kinematic variables are calculated from the energy E'_e and angle θ_e of the scattered electron (measured with respect to the proton direction):

$$y_e = 1 - \frac{E'_e}{E_e} \sin^2 \frac{\theta_e}{2}, \tag{1.6a}$$

$$Q^2_e = 4E'_e E_e \, \cos^2 \frac{\theta_e}{2} = \frac{E'^2_e \sin^2 \theta_e}{1 - y_e} = \frac{p^2_{Te}}{1 - y_e}. \tag{1.6b}$$

The hadron method. The kinematics is measured entirely with the hadronic system:

$$y_h = \frac{E_H - p_{zH}}{2E_e} \qquad Q^2_h = \frac{p^2_{xH} + p^2_{yH}}{1 - y_h}. \tag{1.7}$$

Here E_H, p_{xH}, p_{yH} and p_{zH} denote the 4-vector components of the hadronic system H, which are calculated as the 4-momentum sum over all final state hadrons h. Jacquet and Blondel [9] have shown that the contribution from hadrons lost in the beam pipe is insignificant.

The Sigma method.[10] Here the denominator of y_h is replaced by $\sum_i(E_i - p_{zi})$, where i runs over all final state particles, *including* the scattered electron. This expression equals $2E_e$ due to energy momentum conservation. In case the incident electron had radiated off photons which escape detection, the sum yields the true electron energy which goes into the ep interaction. This method relies on both electron and hadron measurements. With

$$\Sigma = \sum_h (E_h - p_{zh}), \tag{1.8}$$

where the sum runs over all hadronic final state particles, the kinematic variables can be written as

$$y_\Sigma = \frac{\Sigma}{\Sigma + E'_e(1 - \cos\theta_e)} \qquad Q^2_\Sigma = \frac{E'^2_e \sin^2 \theta_e}{1 - y_\Sigma}. \tag{1.9}$$

The denominator $\Sigma + E'_e(1 - \cos\theta_e)$ is twice the energy of the "true" incident electron, after QED radiation from the incoming electron beam.

The double angle method. We define the angle γ by

$$\cos\gamma = \frac{p^2_{TH} - (E_H - p_{zH})^2}{p^2_{TH} + (E_H - p_{zH})^2}. \tag{1.10}$$

In the simple quark parton model γ would be the angle of the scattered (massless) quark. The kinematic variables can be calculated from γ regardless of its interpretation:

$$Q^2_{\mathrm{DA}} = 4E_e^2 \frac{\sin\gamma(1+\cos\gamma)}{\sin\gamma + \sin\theta_e - \sin(\theta_e + \gamma)} \qquad (1.11)$$

and

$$x_{\mathrm{DA}} = \left(\frac{E_e}{E_p}\right) \frac{\sin\gamma + \sin\theta_e + \sin(\theta_e + \gamma)}{\sin\gamma + \sin\theta_e - \sin(\theta_e + \gamma)}. \qquad (1.12)$$

For most of the phase space the electron method is superior. At small y the hadron method has a better resolution than the electron method. The "mixed method" uses Q^2 reconstructed from the electron method and y reconstructed with the hadron method. Also the double angle method and the sigma method use information from both the electron and the hadronic system, thus interpolating between the pure electron and hadron methods. The sigma method has the advantage that it corrects for initial state radiation.

2. Theoretical Framework

2.1 Deep Inelastic Scattering

The fundamental measurement in DIS concerns the cross-section for $ep \rightarrow e'H$ as a function of the kinematic variables (any pair of two independent ones). The quark parton model (QPM) offers a physical picture: the scattering takes place via a virtual photon which is radiated off the scattering electron, and which couples to a pointlike constituent inside the proton, that is a quark or antiquark. The cross-section is then proportional to the quark density inside the proton.

The differential cross-section $ep \rightarrow e'H$ can be expressed in terms of two[1] independent structure functions $F_1(x, Q^2)$ and $F_2(x, Q^2)$:

$$\frac{d^2\sigma}{dx dQ^2} = \frac{4\pi\alpha^2}{xQ^4} \left[(1-y)F_2 + y^2 x F_1 \right]$$
$$= \frac{4\pi\alpha^2}{xQ^4} \left[1 - y + \frac{y^2}{2} \frac{1}{1+R} \right] F_2 \qquad (2.1)$$
$$= \frac{4\pi\alpha^2}{xQ^4} \left[\left(1 - y + \frac{y^2}{2} \right) \cdot F_2 - \frac{y^2}{2} \cdot F_L \right].$$

α is the electromagnetic coupling constant. Here we have expressed the cross-section also in terms of the longitudinal structure function $F_L(x, Q^2)$ and the ratio R, defined as

$$F_L := F_2 - 2x F_1 \qquad (2.2)$$

and

$$R := \frac{F_L}{F_2 - F_L} = \frac{F_2 - 2x F_1}{2x F_1} = \frac{\sigma_L}{\sigma_T}. \qquad (2.3)$$

R can be interpreted as the ratio of the cross-sections σ_T and σ_L for the absorption of transversely and longitudinally polarized virtual photons on protons, with $\sigma_{tot}^{\gamma^* p} = \sigma_L + \sigma_T$. The structure function F_2 can be expressed[2] in terms of σ_T and σ_L,

---

[1] For the sake of simplicity, Z exchange (a 1% correction for $Q^2 \approx 1000$ GeV^2) has been neglected, and the structure function F_3 has thus been omitted.

[2] We use the Hand convention [11] for the definition of the virtual photon flux.

$$F_2 = \frac{Q^2(1-x)}{4\pi^2\alpha} \frac{Q^2}{Q^2 + 4m_p^2 x^2} \cdot \sigma_{\text{tot}}^{\gamma^* p} \approx \frac{Q^2}{4\pi^2\alpha}(\sigma_L + \sigma_T), \tag{2.4}$$

where the small x approximation has been applied. Similarly,

$$F_L = \frac{Q^2(1-x)}{4\pi^2\alpha} \cdot \sigma_L \approx \frac{Q^2}{4\pi^2\alpha} \cdot \sigma_L. \tag{2.5}$$

In the "DIS" scheme F_2 can be written in terms of the quark and anti-quark densities, q_i and $\overline{q}_i$, and their couplings to the photon, i.e. their charges e_{q_i}:

$$F_2(x, Q^2) = x \sum_i e_{q_i}^2 \left[q_i(x, Q^2) + \overline{q}_i(x, Q^2) \right], \tag{2.6}$$

where the sum runs over all quark flavours.[3] In other schemes (for example the "$\overline{\text{MS}}$" scheme) the relation between F_2 and the parton densities (2.6) holds only in leading order perturbation theory. The longitudinal structure function F_L vanishes in zeroth order α_s, and will be discussed in Sect. 3.3.

In the simple quark parton model the proton consists just of three valence quarks. Their distribution functions in fractional proton momentum x, $xq(x)$, would peak at $x \approx 1/3$ and tend towards zero for $x \to 0, 1$. In a static model of the proton, they would not depend on Q^2. It follows that F_2 should not depend on Q^2, just on x (Bjorken scaling).

When QCD is "turned on" the quarks may radiate (and absorb) gluons, which in turn may split into quark–antiquark pairs or gluon pairs. More and more of these fluctuations can be resolved with increasingly shorter wavelength of the photonic probe, $\lambda = 1/Q$. With Q^2 increasing, we have a depletion of quarks at large x, and a corresponding accumulation at lower x. In addition, "sea quarks" from $g \to q\overline{q}$ splittings populate small x. In fact, at small x it is the gluon content with distribution function $g(x, Q^2)$ that governs the proton and gives rise to the DIS cross-section via the creation of $q\overline{q}$ pairs.

---

[3] Equation (2.6) represents the "leading twist" (called twist 2) contribution to the structure function F_2, when expanded in powers of Q^2 [2],

$$F_2(x, Q^2) = \sum_{n=0}^{\infty} C_n(x, Q^2)/(Q^2)^n.$$

The coefficients $C_n(x, Q^2)$ vary logarithmically with Q^2. The "higher twist" terms ($n > 0$, called twist 4, 6, etc.) arise from interactions of the struck parton with the remnant and are suppressed by $(1/Q^2)^n$. We shall not pursue higher twist effects any further, but note that they may not be negligible at small x for Q^2 as large as a few GeV^2 [2, 12].

2.2 Evolution Equations

It has not yet been possible to calculate the structure of the hadrons from first principles, involving the building blocks of hadronic matter, the quarks and gluons, and their mutual interactions as given by QCD. Therefore also the lepton–nucleon scattering cross-section cannot be calculated from first principles. Due to the factorization theorem of QCD one can however split the problem. The cross-section can be calculated by folding initial parton distribution functions $f_{i/p}$, giving the density of partons i in the proton p, with a perturbatively calculable lepton–parton scattering cross-section. Symbolically

$$\sigma_{ep} = \sum_i \left[f_{i/p} \otimes \sigma_{ei} \right] \tag{2.7}$$

for ep scattering (see Fig.2.1a). The initial parton distributions cannot be calculated. They have to be determined experimentally. They are however universal in the sense that once they have been measured in one reaction, they can be used for calculations of other processes. Rather than employing the rigorous operator product expansion (OPE) technique for the evolution of the structure functions (see e.g. [13]), we shall use in this section the more intuitive picture of Feynman diagram summation.

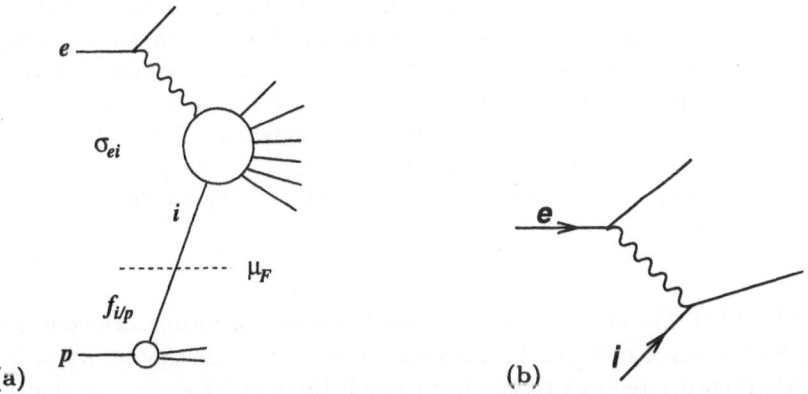

(a)   (b)

Fig. 2.1. (a) Deep inelastic ep scattering. The ep cross-section is factorized into electron–parton cross-sections σ_{ei} and parton densities $f_{i/p}$ with the factorization scale μ_F: $\sigma_{ep} = \sum_i [f_{i/p}(\mu_F^2) \otimes \sigma_{ei}(\mu_F^2)]$. (b) The lowest order diagram (Born graph) contributing to σ_{ei} in (a).

The expansion parameter for the perturbation series is the strong coupling α_s. The coupling is scale dependent according to the "renormalization group equation" (see [14] for a concise summary). The two-loop expression (next-to-leading order = NLO) for the "running" coupling α_s as a function of the renormalization scale μ_R is

$$\alpha_s(\mu_R{}^2) = \frac{4\pi}{\beta_0 \ln(\mu_R{}^2/\Lambda^2)} \left[1 - \frac{2\beta_1}{\beta_0^2} \frac{\ln[\ln(\mu_R{}^2/\Lambda^2)]}{\ln(\mu_R{}^2/\Lambda^2)} \right] \tag{2.8}$$

where

$$\beta_0 = 11 - \frac{2}{3}n_f \qquad\qquad \beta_1 = 51 - \frac{19}{3}n_f. \tag{2.9}$$

The renormalization scale μ_R is set by the length scale $(1/\mu_R)$ over which the interaction takes place, given for example by the virtuality of the probing photon Q^2 in DIS, or by the p_T of a parton. By (2.8) the QCD scale parameter Λ is introduced. In contrast to α_s, its definition depends on the number of active flavours n_f and on the renormalization scheme (for example the "minimal subtraction scheme" $\overline{MS}$). The strong coupling decreases with increasing scale – at short distances partons become asymptotically free. α_s grows beyond all bounds for small scales ($\mu_R \to \Lambda$) or large distances, when perturbation theory breaks down and confinement sets in. The size of a hadron ≈ 1 fm provides an estimate of when that happens. Therefore $\Lambda = \mathcal{O}(1/\text{fm}) \approx 0.2$ GeV. The current world average for α_s at the scale set by the Z mass is $\alpha_s(m_Z^2) = 0.118 \pm 0.003$ or $\Lambda_{\overline{MS}}^{(5)} = 0.209^{+0.039}_{-0.033}$ GeV for five flavours [14].

The calculation of the cross-section σ_{ei} is a formidable task. It turns out that there is no fast convergence of the perturbation series (Fig. 2.2). Many diagrams contribute and have to be summed up. One encounters two types of divergencies. Divergencies due to the radiation of soft quanta with small momenta $k \to 0$ are exactly cancelled by virtual corrections to graphs where that radiation is absent ("no emission"). Divergencies due to collinear radiation (so called collinear or mass singularities for $k_T \to 0$) can be absorbed (factorized off) into the "bare" parton distribution functions. Thereby they are redefined and depend now on the (mass) factorization scale μ_F, and so does the electron–parton scattering cross-section with the singularities removed:

$$\sigma_{ep} = \sum_i [f_{i/p}(\mu_F^2) \otimes \sigma_{ei}(\mu_F^2)] \tag{2.10}$$

(see Fig. 2.1a). The choice of the factorization scale is arbitrary. The physical cross-section σ_{ep} is of course independent of μ_F. Often one chooses $\mu_F^2 = Q^2$, because then σ_{ei} reduces to the Born graph (here $ei \to ei$ by γ exchange, Fig. 2.1b). In the parton picture $f_{i/p}(Q^2)$ is then interpreted as the parton density in the proton as seen by a photon with virtuality (resolving power) Q^2.

Since $\sigma_{ei}(\mu_F^2)$ can in principle be calculated perturbatively for any scale, one can also calculate the change of the redefined parton distribution function with a change of scale. These are the evolution equations. Once a parton distribution function (or structure function) is known at one scale, it can be calculated for any other scale. For the derivation of the evolution equations, one has to perform the perturbative calculation of σ_{ei}, taking into account all contributing graphs (Fig. 2.2). In order to carry out the calculation in

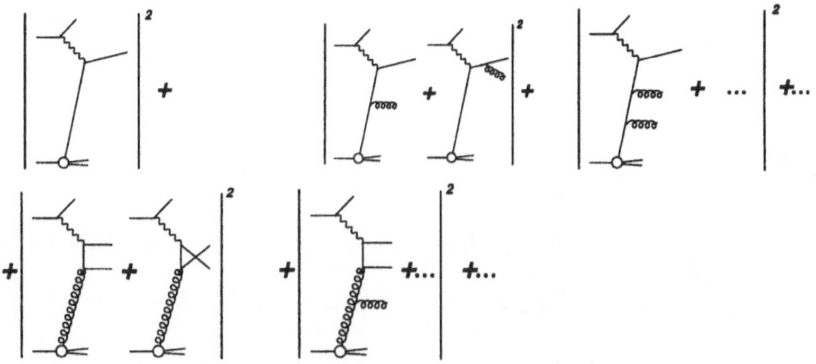

Fig. 2.2. Diagrams contributing to the DIS cross-section σ_{ep}

practice, one applies certain approximations, thus restricting the phase space for radiation. Such approximations are then valid in regions of x and Q^2 where the selected contributions are the dominant ones. In the following, evolution equations will be discussed which differ in their approximations, and therefore in their x, Q^2 regions of validity. Always, in order to allow perturbation theory to be valid, $\alpha_s(Q^2) \ll 1$ is required.

In a "physical" gauge, in which only the physical transverse gluon polarization states need to be taken into account, the individual contributions of the perturbation series can be represented by so-called ladder diagrams (see Fig. 2.3). We work in a frame that moves parallel to the proton, and where the proton is fast. The transverse momenta of the emitted quanta are denoted with p_{Ti}. Similarly, the transverse momenta carried by the quanta that constitute the side rails of the ladder are k_{Ti}. The longitudinal components are given in fractions of the proton energy E_i/E_p and are labelled with ξ_i for the emitted quanta and with x_i for the internal quanta. Energy–momentum conservation requires $x_i = x_{i+1} + \xi_i$, and therefore $x_i > x_{i+1}$.

2.3 The DGLAP Equations

In the approximation leading to the DGLAP (Dokshitzer–Gribov–Lipatov–Altarelli–Parisi) equations [15] all ladder diagrams are summed up, in which the transverse momenta along the side rails of the ladder are "strongly ordered", $Q_0^2 \ll \cdots \ll k_{Ti}^2 \ll k_{Ti+1}^2 \ll \cdots \ll Q^2$. This condition implies strong ordering also for the emitted quanta, $p_{Ti} \ll p_{Ti+1}$.

Where can we expect such an approximation to be valid? A good pedagogical discussion can be found in [16]. We shall sketch the main arguments. The evaluation of a ladder diagram with n rungs requires integrations over the internal momenta exchanged between rungs of the form

$$\alpha_s \cdot \int \frac{\mathrm{d}k_{Ti}^2}{k_{Ti}^2} \cdots \int \frac{\mathrm{d}x_i}{x_i} \cdots, \tag{2.11}$$

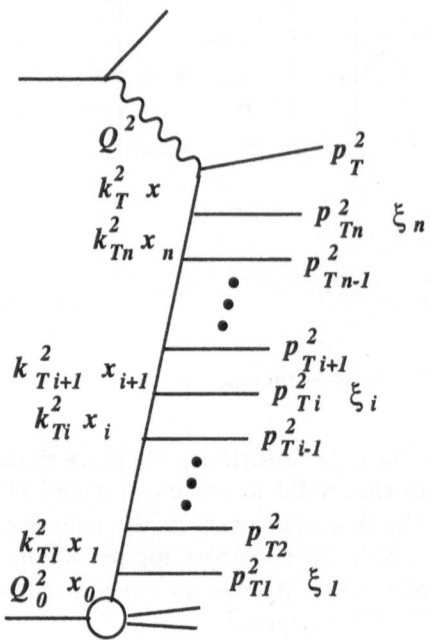

Fig. 2.3. The notation for a ladder diagram with n rungs

where the dots represent functions which depend on the actual nature of the emitted quanta and their dynamics. With strong k_T ordering, the nested integration over all n rungs in the ladder can be carried out. The result is an expression $\propto (\alpha_s \ln(Q^2/Q_0^2))^n$. We see that the k_T integration yields large logarithms when the k_T are strongly ordered. They compensate the smallness of α_s. Clearly, since α_s decreases only logarithmically with Q^2 and is compensated by a logarithmically growing term in Q^2 , in a perturbative expansion all graphs with rungs up to $n = \infty$ need to be summed up. (Often the expression "resummation" is used, because one rearranges the perturbation series such that the largest terms come first.) This is called a leading log approximation (here in $\ln(Q^2/Q_0^2)$), since each power n in α_s is accompanied by the same (maximal) power of $\ln(Q^2/Q_0^2)$. Subleading terms would be $\propto \alpha_s^n (\ln(Q^2/Q_0^2))^{n-1}$.

We expect this to be a good approximation when Q^2 is large, but x is not too small in order not to produce also large logarithms:

$$\alpha_s(Q^2) \ln \frac{1}{x} \ll \alpha_s(Q^2) \ln \frac{Q^2}{Q_0^2} \lesssim 1. \tag{2.12}$$

In this approximation, the evolution equations for the quark density q_i for flavour i and the gluon density g are

$$\frac{dq_i(x, Q^2)}{d \ln Q^2} = \frac{\alpha_s}{2\pi} \int_x^1 \frac{dz}{z} \left[q_i(z, Q^2) P_{qq} \left(\frac{x}{z} \right) + g(z, Q^2) P_{qg} \left(\frac{x}{z} \right) \right]$$

$$\frac{dg(x,Q^2)}{d\ln Q^2} = \frac{\alpha_s}{2\pi} \int_x^1 \frac{dz}{z} \left[\sum_i q_i(z,Q^2) P_{gq}\left(\frac{x}{z}\right) + g(z,Q^2) P_{gg}\left(\frac{x}{z}\right) \right].$$
(2.13)

These are the famous DGLAP equations [15], describing the scaling violations of the structure functions. They involve the calculable Altarelli–Parisi
splitting functions $P_{ij}(\zeta)$. $(\alpha_s/2\pi) P_{ij}(\zeta)$ gives the probability per unit of
$\ln(Q^2/Q_0^2)$ for parton branchings $q \to qg$, $g \to gg$ and $g \to q\bar{q}$, where the
daughter parton i carries a fraction $1 - \zeta$ of the mother's (j) momentum.
The splitting functions in LO are given by

$$
\begin{aligned}
P_{qq}(\zeta) &= \tfrac{4}{3}\tfrac{1+\zeta^2}{1-\zeta} = P_{gg}(1-\zeta) \\
P_{gq}(\zeta) &= \tfrac{4}{3}\tfrac{1+(1-\zeta)^2}{\zeta} = P_{qq}(1-\zeta) \\
P_{qg}(\zeta) &= \tfrac{1}{2}(\zeta^2 + (1-\zeta)^2) = P_{qg}(1-\zeta) \\
P_{gg}(\zeta) &= 6(\tfrac{\zeta}{1-\zeta} + \tfrac{1-\zeta}{\zeta} + \zeta(1-\zeta)) = P_{gg}(1-\zeta).
\end{aligned}
$$
(2.14)

The singularities for soft emissions $\zeta \to 1$ are cancelled by virtual corrections
to the "no-emission" graphs. This is physical, because arbitrary soft emissions
cannot be distinguished from the "no-emission" case.[4] The coupled integro-
differential equations for the quark and gluon densities (2.13) can be solved,
allowing one to calculate them for any value of Q^2 and $x > x_0$, once they are
known at a particular value Q_0^2 for $x > x_0$.

A special case for which the DGLAP equations can be solved analytically
(see for example [13]) occurs when in addition to the above conditions also
strong ordering in x is required, $x \ll ... \ll x_{i+1} \ll x_i \ll ... \ll x_0$. The
large logarithmic terms arising from the integration are then of the form
$\propto (\alpha_s(Q^2) \ln(Q^2/Q_0^2) \ln(1/x))^n$, which need to be resummed. This is the
double leading log approximation (DLL). It is expected to hold when the
DLL terms dominate over the others:

$$
\left.\begin{aligned}
\alpha_s(Q^2)\ln\frac{Q^2}{Q_0^2} \\
\alpha_s(Q^2)\ln\frac{1}{x}
\end{aligned}\right\} \ll \alpha_s(Q^2)\ln\frac{Q^2}{Q_0^2}\ln\frac{1}{x} \lesssim 1.
$$
(2.15)

This is the case for large Q^2 and small x. At small x the parton content of
the proton is expected to be dominated by gluons, because $P_{ij}(\zeta = x/z)$ is
largest when gluons are being produced ($i = g$). When quarks are neglected,
and $P_{gg} = 6/\zeta$ is approximated, the DGLAP equations can be solved to yield
the DLL solution [17]

---

[4] Technically the singularities can be regularized by the following prescription.
Replace $1/(1 - \zeta)$ with $1/(1 - \zeta)_+$, where the "+ prescription" defines how
integrals are to be carried out:

$$\int_0^1 d\zeta \frac{f(\zeta)}{(1-\zeta)_+} := \int_0^1 d\zeta \frac{f(\zeta) - f(1)}{1-\zeta}$$

for any function $f(z)$. A term $2\delta(1 - \zeta)$ has to be added to P_{qq}, and a term
$(11/2 - n_f/3)\delta(1 - \zeta)$ to P_{gg}.

$$xg(x,Q^2) \approx xg(x,Q_0^2)\exp\sqrt{\frac{144}{25}\ln\left[\frac{\ln(Q^2/\Lambda^2)}{\ln(Q_0^2/\Lambda^2)}\right]}\ln(1/x), \qquad (2.16)$$

provided the gluon density is not too singular at small x (needs to be quantified). At small x a fast rise of the gluon density with decreasing x is predicted. That is, xg increases faster than $(\ln(1/x))^\lambda$, but slower than $(\frac{1}{x})^{\lambda'}$ for any powers $\lambda, \lambda' > 0$. Apart from these shape restrictions the actual rate of growth is not predicted; it depends on the "evolution length" from Q_0^2 to Q^2.

2.4 The BFKL Equation

When x is small, but Q^2 is not large enough to reach the DLL regime, the DGLAP approximations cease to be valid. For the limit $1/x$ large and Q^2 finite and fixed the BFKL (Balitsky–Fadin–Kuraev–Lipatov) [18] equation has been derived. It takes into account diagrams in which the x_i are strongly ordered, $x_0 \ll \cdots \ll x_i \ll x_{i+1} \ll \cdots \ll x$. No ordering on k_{Ti} is imposed. Large logarithms $\propto (\alpha_s \ln(1/x))^n$ are thus generated that need to be resummed, leading to the leading log approximation in $\ln(1/x)$. The region of validity is

$$\alpha_s(Q^2)\ln\frac{Q^2}{Q_0^2} \ll \alpha_s(Q^2)\ln\frac{1}{x} \lesssim 1. \qquad (2.17)$$

The BFKL equation is expressed in terms of the "unintegrated" gluon density $f(x, k_T^2)$, which is related to the usual gluon density by

$$xg(x,Q^2) = \int_0^{Q^2} \frac{\mathrm{d}k_T^2}{k_T^2} f(x, k_T^2). \qquad (2.18)$$

The BFKL equation is an evolution equation in x. It is formulated for gluons which dominate at small x,

$$\frac{\partial f(x,k_T^2)}{\partial\ln(1/x)} = \frac{3\alpha_s}{\pi}k_T^2 \int_0^\infty \frac{\mathrm{d}k_T'^2}{k_T'^2}\left[\frac{f(x,k_T'^2)-f(x,k_T^2)}{|k_T'^2-k_T^2|} + \frac{f(x,k_T^2)}{\sqrt{4k_T'^4+k_T^4}}\right]$$

$$= K \otimes f, \qquad (2.19)$$

where K is the BFKL kernel. $f(x, k_T^2)$ can be calculated for any (small) x, once it is known at some x_0 for all k_T^2. We note in passing that if one requires strong k_T ordering ($k_T^2 \gg k_T'^2$) when solving (2.19), the DLL result is retained [16].

For fixed α_s the equation can be solved analytically. The result is (in the saddle point approximation)

$$\frac{f(x,k_T^2)}{\sqrt{k_T^2}} \propto \frac{(x/x_0)^{-\lambda}}{\sqrt{2\pi\lambda''\ln(x_0/x)}}\exp\left[\frac{-\ln^2(k_T^2/\overline{k}_T^2)}{2\lambda''\ln(x_0/x)}\right]$$

$$\propto \left(\frac{x}{x_0}\right)^{-\lambda} \qquad (2.20)$$

with $\lambda = (n_c \alpha_s / \pi) \cdot 4 \ln 2 \approx 0.5$ (for $n_c = 3$ colours and $\alpha_s = 0.19$), and $\lambda'' = (n_c \alpha_s / \pi) \cdot 28 \zeta(3)$ (the Riemann ζ function gives $\zeta(3) = 1.202$). $\bar{k}_T^2$ specifies the starting point for the evolution. Therefore the gluon density is expected to rise like a power of $1/x$ for decreasing x, $xg(x, Q^2) \propto x^{-\lambda}$, faster than the DLL result (2.16). However, the running of α_s and higher-order corrections decrease the value of $\lambda \approx 0.5$ [19].

Another characteristic prediction of the BFKL equation is "k_T diffusion", in contrast to k_T ordering for DGLAP (see Fig. 2.4). The k_T distribution function $f(x, k_T^2)/\sqrt{k_T^2}$ is Gaussian in $\ln k_T^2$ with a width that increases with the BFKL "evolution length" $\sqrt{\ln(x_0/x)}$. An individual evolution path will follow a kind of random walk in k_T^2. An ensemble of evolution paths exhibits a diffusion pattern according to the Gaussian in $\ln k_T^2$.

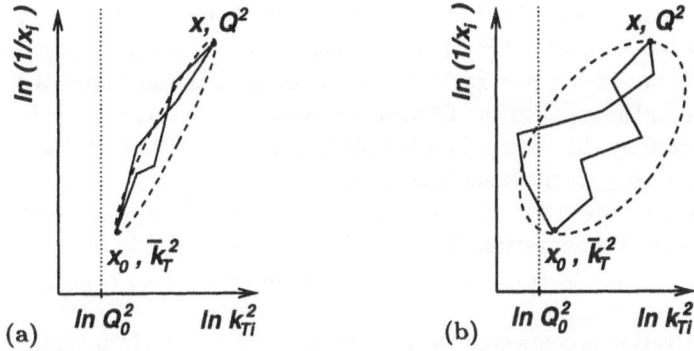

(a)   (b)

Fig. 2.4. Possible evolution paths for the parton cascade **(a)** for DGLAP evolution with strong k_T ordering and **(b)** for BFKL evolution without k_T ordering. Shown are the x_i and k_{Ti} in the ladder for fixed start $(x_0, \bar{k}_T^2)$ and end (x, Q^2) points. The infrared region $k_{Ti}^2 < Q_0^2$ is to the left of the dotted line. The dashed "cigars" represent the $\pm 1\sigma$ contours for the (x_i, k_{Ti}) taken from a large ensemble of evolution paths. One of the BFKL paths shown "diffuses" into the infrared region $k_{Ti}^2 < Q_0^2$

k_T diffusion poses a difficulty for the application of the BFKL equation, because k_T may diffuse into the infrared region ($k_T < Q_0$) where perturbation theory cannot be applied. One therefore usually introduces a lower cut-off k_{T0} for the k_T integration, and studies the dependence of the result on that cut-off. Due to k_T diffusion BFKL loses much of its predictive power when applied to the structure function F_2. The inclusive structure function F_2 is probably not a good place to identify BFKL effects unambiguously. It is however possible to study special final state configurations where diffusion into the infrared region can be minimized by fixing both the start and end point of the evolution far enough above the infrared region. The search for signs of BFKL evolution in the hadronic final state is presented in Chap. 8.

The CCFM equation [20] developed in recent years unifies the BFKL and DGLAP approaches [21] and takes into account coherence effects by angular ordering. The CCFM approach leads to a reduction of the exponent λ and a reduction of the k_T diffusion [22, 23]. The linked dipole chain model [24] provides an implementation of the CCFM equation which is suited for final state predictions.

2.5 The Interest in Small x

Orthogonal Evolution Equations

In Fig. 2.5 the regions of validity of the different evolution equations are sketched. They are not predicted precisely by the theory, but have to be explored experimentally.

With DGLAP evolution a parton density $p(x, Q_0^2)$ known for $x \in [x_0, 1]$ can be evolved to any value of Q^2 for $x \in [x_0, 1]$. The behaviour for $x < x_0$ cannot be predicted with DGLAP. Similarly, with BFKL evolution a parton density $p(x_0, k_T^2)$ known for $0 < k_T^2 < \infty$ can be evolved to any value of x. The new feature, orthogonal to the DGLAP evolution, is that the low x behaviour is predicted by the theory. In principle DGLAP and BFKL evolution together could be used for a Münchhausen trick (bootstrapping), to predict the structure of the proton for all x as long as $Q^2 > Q_0^2$, above a cut-off to avoid the non-perturbative region. The problem with k_T diffusion into the non-perturbative regime however poses a severe obstacle for that goal to be reached.

Another motivation is connected with the cross-section for hadron–hadron scattering at high energies. Before we can make the point, we need to make an excursion to Regge theory (see for example [25] as a review related to

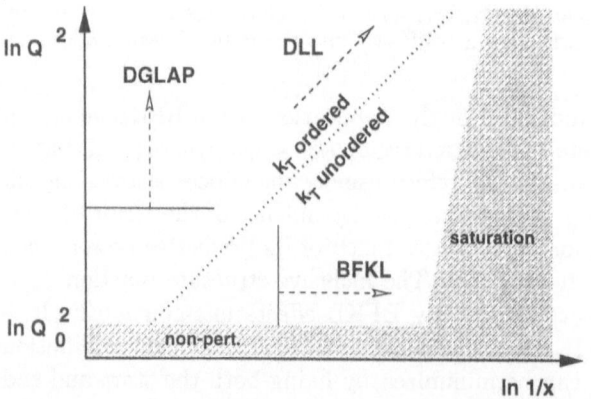

Fig. 2.5. Schematic map of the kinematic plane. Indicated are the actions and regions of validity of the DGLAP, DLL and BFKL evolutions. The non-perturbative region ($Q^2 < Q_0^2$) and the saturation limit are given by the shaded areas

HERA physics; [26] and [27] as a modern language textbook introduction; [28] for an in depth discussion of Regge theory).

Regge Theory

Consider the elastic scattering of hadrons a and b, $ab \rightarrow ab$ (Fig. 2.6a). Their 4-momenta are denoted by a, b for the initial and a', b' for the final state. The cross-section can be expressed as a function of the Mandelstam variables $s = (a + b)^2$ and $t = (b - b')^2$. It is the squared sum over the scattering amplitudes due to the quanta X (conventionally mesons) that can be exchanged,

$$\frac{d\sigma^{\text{el.}}}{dt} = \frac{1}{16\pi s^2} \left| \sum_X \mathcal{M}_X^{\text{el.}}(s, t) \right|^2 . \tag{2.21}$$

In Regge theory, where the mesons are connected via a so-called Regge trajectory (explained below), the sum yields

$$\sum_X \mathcal{M}_X^{\text{el.}}(s, t) = \mathcal{M}_{IR}(s, t) \propto \beta(t)\xi(\alpha(t))s^{\alpha(t)}. \tag{2.22}$$

$\beta(t)$ is an unknown real function. The complex phase is given by

$$\xi(\alpha(t)) = e^{-i\pi\alpha(t)/2} \tag{2.23}$$

for exchanged particles with C parity $+1$, and receives an extra factor i for negative C parity. The Regge trajectory $\alpha(t)$ gives the relationship between the mass m and the spin J of the exchanged mesons, $J = \alpha(m^2)$ (Fig. 2.6b).

Empirically, Regge trajectories can be parametrized as straight lines with

$$\alpha(t) = \alpha_0 + \alpha' t. \tag{2.24}$$

α_0 is called the intercept (with the ordinate), and α' the slope of the trajectory. As an example, a Regge trajectory for mesons is shown in Fig. 2.6b. Due to the confinement problem, meson trajectories (hadron masses) could not yet be calculated from first principles in QCD.

The Total Cross-Section. Starting from the elastic cross-section (in principle one has to sum over all Regge trajectories whose resonances can be exchanged in the reaction)

$$\frac{d\sigma^{\text{el.}}}{dt} \propto (\beta(t))^2 \cdot s^{2\alpha(t)-2}, \tag{2.25}$$

we can use the optical theorem, relating the total cross-section to the forward scattering amplitude

$$\sigma_{\text{tot}} = \frac{1}{s} \text{Im} \mathcal{M}_{\text{el}}(s, t = 0), \tag{2.26}$$

to predict the behaviour of the total hadron–hadron scattering cross-section

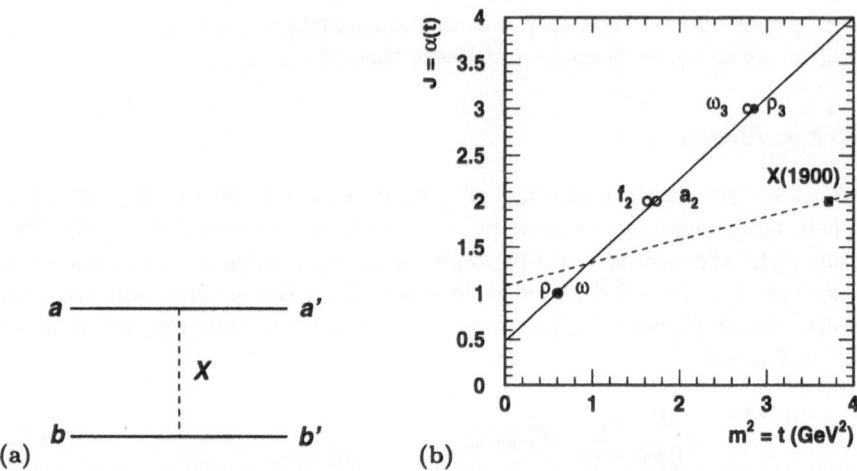

Fig. 2.6. (a) The elastic scattering of hadrons a and b, mediated by the exchange of a quantum X. The 4-vectors before and after the scattering are a, b and a', b'. (b) Regge trajectories. The trajectories for the ρ, ω, f_2 and the a_2 resonances almost coincide; they are represented here with a *solid line*. The indicated resonances are the $\rho(770)$, the $\omega(782)$, the $f_2(1270)$, the $a_2(1320)$, the $\rho_3(1690)$ and the $\omega_3(1670)$. The Pomeron trajectory (*dashed line*) is shown together with the $I(J^{PC}) = 0(2^{++})$ state $X(1900)$ observed by WA91 [29]

$$\sigma_{\text{tot}} \propto s^{\alpha_0 - 1}. \tag{2.27}$$

The total cross-sections for pp, $p\bar{p}$, γp and $\gamma\gamma$ reactions are plotted as a function of the CM energy $\sqrt{s}$ in Fig. 2.7. Their behaviour is surprisingly similar (and also for other hadron–hadron scattering cross-sections like πp, Kp etc. that are not shown in Fig. 2.7 [14]). They fall at small CM energy $\sqrt{s} \lesssim 10$ GeV, and rise towards large energy. All these cross-sections can be parametrized with the universal ansatz [30]

$$\sigma_{\text{tot}} = A s^{\alpha_{IM}(0)-1} + B s^{\alpha_{IP}(0)-1}. \tag{2.28}$$

A and B are process-dependent constants, whereas $\alpha_{IM}(0) = 0.5322 \pm 0.0059$ and $\alpha_{IP}(0) = 1.0790 \pm 0.0011$ [14] are universal, process-independent constants.

The Pomeron

The fall-off at small energies is readily interpreted as being due to meson exchange, whose Regge trajectories have an intercept $\alpha_{IM}(0) < 1$ (compare Fig. 2.6b; here only trajectories that dominate high-energy scattering are shown; other meson trajectories have smaller intercepts and therefore do not contribute much to high-energy scattering). Correspondingly, the rise at high energy is attributed to an exchange described by a Regge trajectory with intercept $\alpha_{IP}(0) > 1$. There exists however no established set of particles with

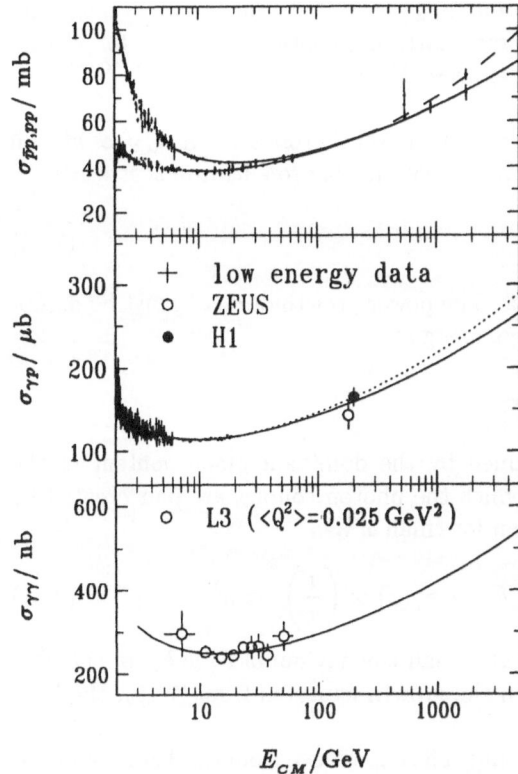

Fig. 2.7. The total cross-section for pp (or $p\bar{p}$), γp and $\gamma\gamma$ interactions as a function of the total CM energy $\sqrt{s} = E_{CM}$ [31]. ZEUS and H1 measure $\sigma_{\gamma p}$ in ep scattering with $Q^2 \approx 0$ (photoproduction). The curves represent the DL parametrizations with $\alpha_{IP} = 1.0808$ (*solid line*), $=1.112$ (*dashed line*) and $=1.088$ (*dotted line*)

such a Regge trajectory. Nevertheless, the Regge ansatz gives a successful parametrization of the scattering process. Therefore a hypothetical object, the Pomeron IP is postulated, which has the quantum numbers of the vacuum (electrically and colour neutral, isospin 0 and C parity +1), and whose exchange is described by the Pomeron trajectory $\alpha_{IP}(t) \approx 1.08 + 0.25 \cdot t$, see Fig 2.6b. The object is suspected to be of gluonic nature, perhaps a glue ball. A possible glue ball candidate with $J^{PC} = 2^{++}$ in fact would fall on the Pomeron trajectory, see Fig. 2.6b. Because $\alpha_{IP}(0) > 1$, physical states with $J = 0, 1$ belonging to the Pomeron trajectory cannot exist. An up-to-date textbook on QCD and the Pomeron is [27].

Deep inelastic scattering at small x can be viewed as virtual photon–proton scattering at high energy $\sqrt{s_{\gamma^* p}} = W \approx \sqrt{Q^2/x}$. We had connected the structure function F_2 with the total cross-section for $\gamma^* p$ scattering,

$$F_2 = \frac{Q^2}{4\pi^2\alpha}\sigma_{tot}^{\gamma^* p}. \tag{2.29}$$

If the total cross-section behaviour $\sigma_{tot} \propto s^{0.08}$ found for hadron–hadron scattering and real photon–hadron scattering continues to hold for virtual photon–hadron scattering, one expects F_2 to rise as $F_2 \propto (1/x)^{0.08}$ with decreasing x.

We note that the power growth of the total cross-section $\sigma_{tot} \propto s^\lambda$ will eventually violate the Froissart bound [32] for hadron–hadron scattering,

$$\sigma_{tot} \leq \frac{\pi}{m_\pi^2} \left(\ln \frac{s}{s_0} \right)^2, \tag{2.30}$$

where s_0 is an unknown constant. The power growth of σ_{tot} must be dampened by some mechanism at large energies.

The Small x Behaviour of F_2

At small x, F_2 will be determined by the dominant gluon content of the proton, because the quarks to which the photon couples are pair created by the gluons. The BFKL prediction for small x was

$$xg(x) \propto \left(\frac{1}{x} \right)^\lambda \Rightarrow \sigma_{tot}^{\gamma^* p} \propto F_2 \propto xg(x) \propto \left(\frac{1}{x} \right)^\lambda \propto s^\lambda. \tag{2.31}$$

The (LO) BFKL expectation for the small x behaviour of F_2 is $F_2 \propto (1/x)^{0.5}$. This power growth is faster than the growth expected from (2.16), the DLL approximation.

This situation is very interesting. The experience from total cross-sections at high energies would suggest that F_2 should rise $\propto (1/x)^{0.08}$ at small x, a behaviour long known and parametrized with the soft Pomeron, but whose origin is not understood from QCD. On the other hand, in the BFKL approximation, QCD does make a prediction for small x, which is different from past experience: F_2 should rise much faster, $\propto (1/x)^{0.5}$. The DLL expectation is in between. The slow rise is often said to be due to the "soft Pomeron" or the "non-perturbative" Pomeron, or the "Donnachie–Landshoff" Pomeron. The fast rise would be attributed to the "hard", or "perturbative", or "Lipatov", or "BFKL" Pomeron, if one still wants to use the Regge language (see Fig. 2.8). Under which conditions will we see which behaviour? How about the transition region? Do HERA data extend into a kinematic regime where the steep rise of F_2 predicted by BFKL can be seen? And if a steep rise is to be seen, is it really to be attributed to BFKL dynamics? Some of the answers will be given by the HERA data presented in Chap. 3 on structure function measurements. It will turn out however that F_2 is too inclusive a quantity to resolve the question of BFKL dynamics. The search for specific signatures of BFKL evolution in the hadronic final state is presented in Chap. 8.

Ultimately the hope is that the BFKL equation offers a way to approach the confinement problem of QCD. It allows us to make predictions for F_2 at small x, thus for the structure of hadrons at small x, something which otherwise has to be assumed as non-perturbative input for the parton densities. It

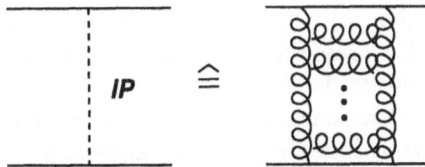

Fig. 2.8. The exchange of a (BFKL) Pomeron in the Regge language is equivalent to a sum over graphs with gluon ladders between the interacting particles in perturbative QCD

also makes a prediction for the total $\gamma^* p$ cross-section at high energies, where previously we had just a prediction based upon a parametrization of "soft" phyics, the "soft" Pomeron. That the two predictions do not coincide makes it all the more interesting.

2.6 Hadronic Final States

So far we have developed the theory for the total inclusive cross-section, loosely speaking a sum over everything that can happen inside the proton. When the proton is being probed by the virtual photon, one out of all the possible virtual fluctuations in the proton is selected by the measurement process. The remnants of the fluctuations materialize in the hadronic final state and become observable when the proton wave function is projected onto a specific state. Also the scattered quark and radiation thereof contribute to the hadronic final state. Matters are even more complicated: the distinction between initial state fluctuations and final state radiation may be practical and justifiable in many cases, but is quantum mechanically not rigorous.

QCD predictions for the hadronic final state are in general much more difficult and less rigorous than for the inclusive cross-section. On the other hand the hadronic final state can provide much more detailed information on the QCD processes in *ep* scattering than just the total cross-section.

Different Observables and Approximations

It depends very much on the final state observable which technique and approximation is appropriate for a theoretical description. For example, for a global event property like the amount of energy on average emitted in a certain solid angle, it may be a good approximation to interpret the parton evolution equations in a probabilistic way.[5] The evolution equations define a

---

[5] Strictly speaking the evolution equations have been derived for the inclusive cross-section. Cancellations between different diagrams are necessary to ensure that the total cross-section stays finite. It is therefore not a priori clear to what extent the evolution equations can be used for hadronic final state predictions.

cascade of parton emissions (a parton shower) with specified emission probabilities. One has to sum up all emission cross-sections, weighted with the energy. Techncially the calculation could be done analytically, numerically, or with a "Monte Carlo" program. Of course such a calculation will only be valid where the approximations that went into the evolution equations are valid.

For less global quantities, one may find that one has selected a piece of the phase space which is unimportant for the total inclusive cross-section, and which might have been rightfully neglected in the evolution equations. For example, for the process $\gamma^* g \rightarrow q\bar{q}$ with high transverse momenta of the quarks, the DGLAP approximation with strong p_T ordering will not be sufficient to describe the cross-section accurately. In that case one will instead use the exact, fixed-order QCD matrix element for the process $\gamma^* g \rightarrow q\bar{q}$ to cover the full phase space, and fold it with the probability to find a gluon in the proton. In practice the matrix element will be calculated only to a few orders in perturbation theory, in leading (LO) or next-to-leading order (NLO).[6] In turn, this approximation will be insufficient for our first example, where many parton emissions contribute. This case is hardly covered by the fixed order matrix element (compare Sect. 5.1); in NLO there are at most three partons in the final state. The actual implementations of the different perturbative QCD approximations are presented in Sect. 4.4.

A special rôle is played by "infrared safe" observables. An infrared safe observable does not change its value when an object in the final state (parton, particle, energy cluster, ...) is split collinearly into smaller units, or when a soft object is added. Examples for infrared safe observables are collective observables like energy flows and certain event shape and jet observables. A counterexample is the total multiplicity. Infrared safety ensures that perturbative predictions can be made without the need to introduce a cut-off parameter against infrared divergencies. This property is also experimentally advantageous, as it does not require us to know the particle composition of a certain energy deposition in the detector that is used in the analysis. For example, calorimeter clusters usually do not have a one-to-one correspondence to incident particles.

Event Generators

The calculational techniques differ. Some specific final state observables can be calculated analytically (rarely), or numerically with a program. More ambitious are event generators, "Monte Carlo" programs (see Sect. 4.7). They attempt to model event by event the complete final state. Ideally, an ensemble of Monte Carlo generated events would correspond in all aspects to an ensemble of events collected in nature. An intrinsic difficulty with this approach

---

[6] Calculations in next-to next-to-leading order (NNLO) are not yet available for DIS.

is that the probabilistic generators have to use probabilities, not probability amplitudes. Quantum mechanical interference effects therefore have to be implemented "by hand", for example by enforcing an angular ordering for parton emissions (see Sect. 4.4).

Hadronization

One of the biggest problems of QCD is to make contact with the real world of hadrons. Perturbative QCD makes predictions for parton cross-sections, while the observable objects are hadrons. The transition from partons to hadrons ("hadronization") cannot be treated perturbatively, because it happens at a scale where the strong coupling constant becomes large. Non-perturbative QCD can be considered as one of the last frontiers, where new insight into nature can be gained, or as a nuisance, because it hides the beauty and simplicity of the partonic world from direct observation.

In any case, much has been learnt about hadronization from the study of hadron production. This has led to some rather successful phenomenological models to describe hadronization (see Sect. 4.5). Being models, they are not uniquely defined by theory. Rather, they depend to some extent on ad hoc assumptions and parameters which are being adapted to observation. At least the hadronization models are universal and can be used for different kinds of reactions. Probably the most sophisticated model is based on a colour string (a flux tube) that connects coloured partons. When stretched by the separating partons, the string breaks, pulling new $q\bar{q}$ pairs from the vacuum. When no more energy is left in the string, colour-neutral hadrons are formed.

3. Inclusive Cross-Sections

3.1 The Structure Function F_2

ZEUS [33, 34, 35, 36] and H1 [37, 38, 39, 40, 41, 42] have measured F_2 in a completely new kinematic domain compared with fixed target experiments, most notably towards much smaller values of x, and towards much larger Q^2. For kinematic reasons the cross-section falls $\propto (1/Q)^4$ (see (2.1)). Here we shall concentrate on the region of Q^2 of less than a few thousand GeV^2, because very little (though not uninteresting [43, 44, 45]) data exist for higher Q^2, especially as the hadronic final state is concerned.

In Fig. 3.1 F_2 is shown as a function of Q^2 for fixed x values. At $x \approx 0.1$ scaling is observed – F_2 does not depend on Q^2. At $x > 0.1$ F_2 decreases with Q^2 due to parton splittings, the products of which are found then at smaller x. Therefore F_2 increases with Q^2 for $x < 0.1$ in accord with the qualitative discussion of Sect. 2.1. When plotted as a function of x for fixed Q^2 (Fig. 3.2), F_2 exhibits a steep rise towards small x, which flattens at smaller Q^2. The increase signals growing parton densities with decreasing x.

From a DGLAP evolution of pre-HERA data this sharp rise could not be predicted a priori, because input distributions at small x were not available. (Assuming however valence-like parton distributions at a small scale $Q_0^2 \approx 0.3$ GeV^2, a sharp rise at larger Q^2 was predicted from DGLAP evolution [49].) It was known though that asymptotically for $Q^2 \rightarrow \infty$ the small x behaviour is given by the DLL formula (2.16). Are the HERA data still consistent with DGLAP evolution, or is there a need for other effects, for example BFKL, which one may expect at very small x? It turns out that the data with $Q^2 \gtrsim 1$ GeV^2 can be fit perfectly well with parton densities which obey the next-to-leading-order (NLO) DGLAP evolution equations (see Fig. 3.2) [34, 39, 41]. Standard QCD evolution appears to work over many orders of magnitude in both x and Q^2!

When fitting $F_2 \propto (1/x)^\lambda$, H1 finds that the exponent λ increases from $\lambda \approx 0.15$ to $\lambda \approx 0.40$ between $Q^2 = 1$ GeV^2 and $Q^2 = 1000$ GeV^2 [40]. F_2 rises faster than expected from the soft Pomeron model ($\lambda \approx 0.08$), but less fast than expected from the LO BFKL equations ($\lambda = 0.5$ in LO; λ is expected to decrease in NLO). In fact, both the x and the Q^2 dependence of F_2 can be attributed predominantly to the DLL formula (2.16), which can be displayed nicely with a suitable variable transformation ("double asymptotic

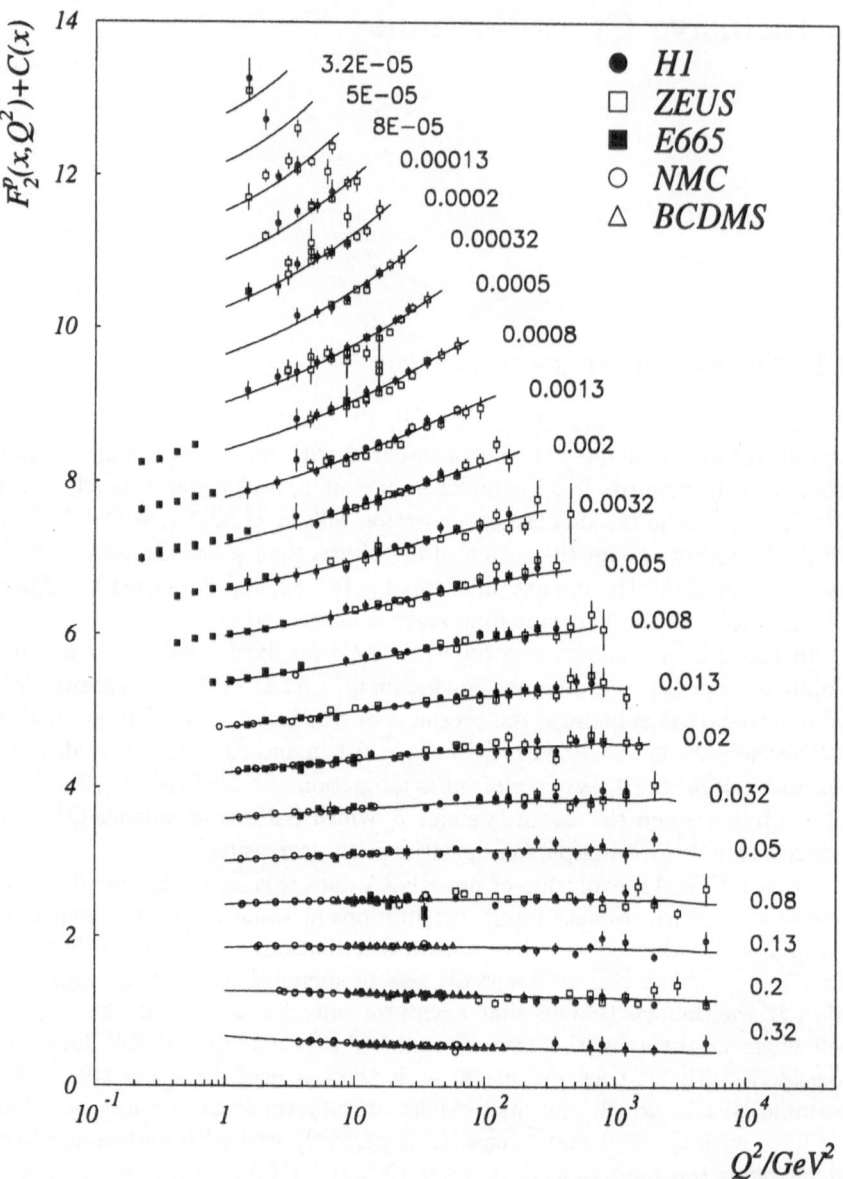

Fig. 3.1. The structure functions $F_2(x, Q^2)$ as a function of Q^2 for different x bins. For better visibility, a function $C(x) = 0.6(i - 0.4)$, with i the x bin number ($i = 1$ for $x = 0.32$), is added to F_2 in the plot. Shown are the HERA [34, 39] and fixed target data [46, 47, 48], as well as a NLO QCD fit [39]

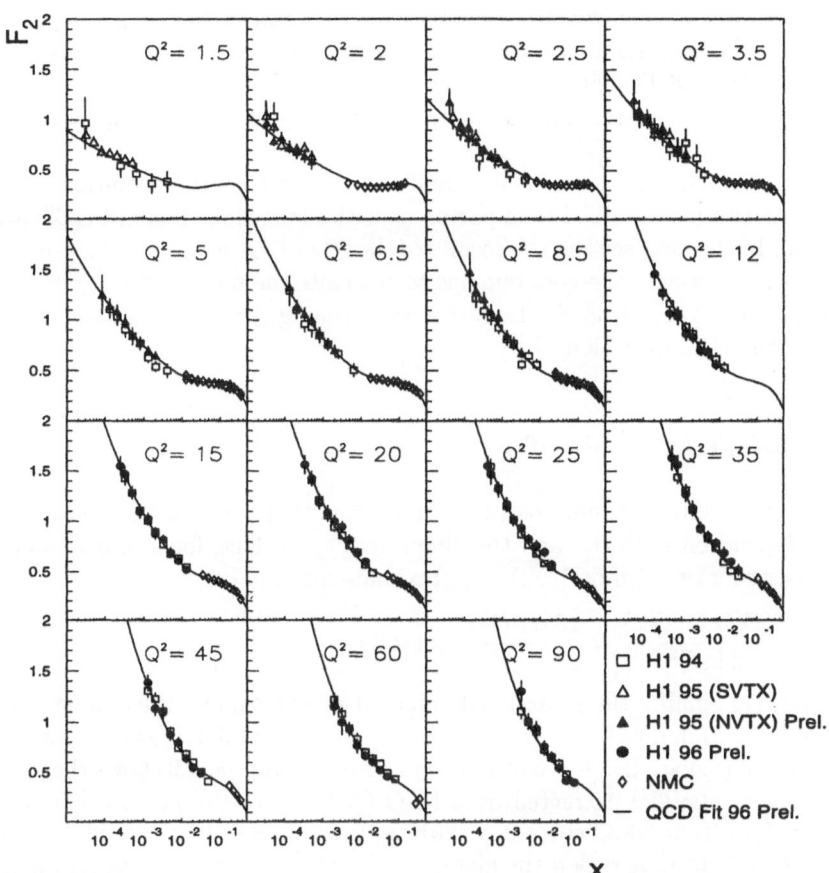

Fig. 3.2. The structure functions $F_2(x, Q^2)$ as a function of x for different Q^2 bins (Q^2 values are given in GeV^2). Shown are data from H1 [41] and NMC [47], together with a NLO DGLAP QCD fit to the data with $Q^2 > 1.5$ GeV^2 from H1, NMC and BCDMS [48]. The starting point for the evolution was $Q_0^2 = 1$ GeV^2, and $\alpha_s(m_Z^2) = 0.118$ was set

scaling" [50]). A unified BFKL and DGLAP description of the F_2 data using the unintegrated gluon distribution (2.18) is also possible, with significant contributions from the $\ln 1/x$ resummation [51]. The structure function data are thus compatible with pure DGLAP evolution, but cannot exclude significant contributions from BFKL evolution [21]. The structure function data are probably too inclusive to resolve the question of non-DGLAP evolution. One has to resort to less inclusive measurements on the hadronic final state, which will be discussed in Chap. 8.

The F_2 data can also be described by evolving flat or valence-like input quark and gluon distributions from a very low scale, $Q_0^2 = 0.35$ GeV^2 up in Q^2 as was shown by Glück, Reya and Vogt (GRV) [49]. The steep rise with decreasing x is achieved by the long evolution length from Q_0^2 to Q^2 (see (2.16)). The success of the GRV prediction came as a surprise for many, as perturbative QCD should only be applicable for Q not too close to $\Lambda_{QCD} \approx 0.2$ GeV because otherwise $\alpha_s\,(Q^2) = 12\pi/[(33 - 2n_f)\ln(Q^2/\Lambda_{QCD}^2)]$ (LO) diverges.

To conclude this section, it is quite satisfying that the evolution of the F_2 data can be described with parton densities following standard QCD evolution. In the next section the results of the NLO DGLAP QCD analyses will be given. However, the goal remains to calculate the measured magnitude of the growth $(1/x)^\lambda$ from QCD, rather than tuning it with the starting point Q_0^2 of the DLL evolution.

3.2 QCD Analysis of F_2

From the DGLAP equations (2.13) it is clear that the scaling violations of F_2 depend on both α_s and the gluon density. In fact, for $x < 0.01$ and in lowest order one can derive the approximate formula [52]

$$\frac{dF_2(x/2, Q^2\ )}{d\ln Q^2} \approx \frac{10}{27}\frac{\alpha_s(Q^2)}{\pi}xg(x, Q^2), \tag{3.1}$$

because at small x the proton is dominated by gluons, and the scaling violations arise from quark pair creation from gluons. The full NLO QCD analyses now employed at HERA are of course more involved. Fig. 3.3 shows the gluon density $x \cdot g(x, Q^2)$ extracted from NLO QCD fits to the F_2 data [53]. Previous data from NMC cover $x > 0.01$, and the HERA data extend down to $x = 0.0001$. In that region the gluon density increases sharply towards small x.

Clearly, the density cannot increase forever; eventually saturation effects will set in [58, 59]. The effective transverse size of a gluon that is probed by a photon of virtuality Q is $\sim 1/Q$. Fluctuations below that scale happen, but cannot be resolved. When such gluons of transverse size $\sim 1/Q$ fill up the whole transverse area offered by the proton, they will start to overlap and recombine. This would be a novel and very interesting situation indeed: high parton density, but Q^2 large enough for $\alpha_s(Q^2)$ to be small! Given the size of the proton ≈ 1 fm, the critical condition for saturation effects to turn on can be estimated [59] as $x_{crit}g(x_{crit}, Q^2) \approx 1\mathrm{fm}^2/(1/Q^2) \approx 25Q^2/\mathrm{GeV}^2$. This value is by far not reached by the measured gluon density (Fig. 3.3a). It could be however that saturation does not set in uniformly over the proton's transverse area, but starts locally in so-called hot spots [60]. In this case x_{crit} would be larger. The inclusive structure function data however do not require any saturation correction. Even a conspiracy of two new effects, a sharp rise of F_2 from BFKL evolution dampened by saturation effects, cannot be ruled out. Probably saturation effects will be first seen in hadronic final state data [60].

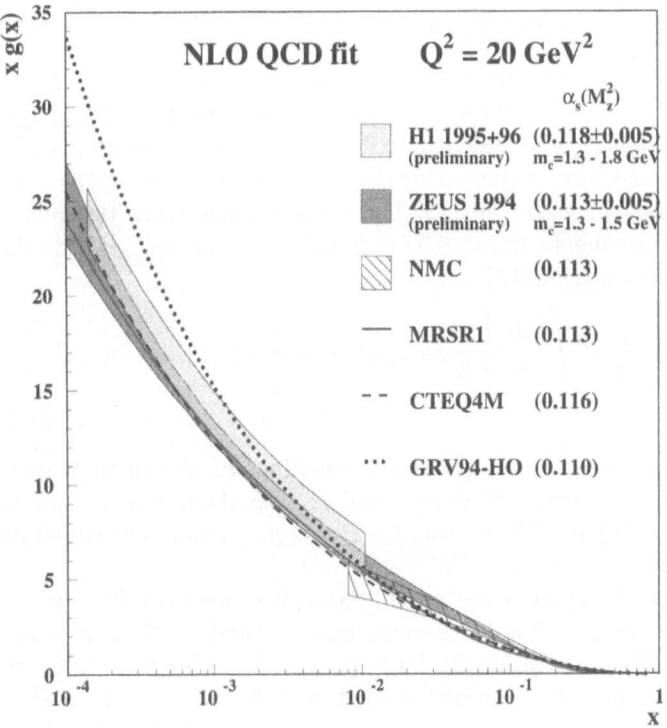

Fig. 3.3. The gluon density $xg(x, Q^2)$ at $Q^2 = 20$ GeV^2 [53], as extracted from NLO QCD analyses of the H1 [41] and ZEUS [54] F_2 data together with the NMC result [55]. Shown are the $\pm 1\sigma$ error bands of the analyses. The values for the charm mass and for $\alpha_s(m_Z^2)$ which were used in the analyses are given. The extracted gluon density is compared to the global analyses MRSR1 [56], CTEQ4m [57] and GRV94-HO [49]

From an analysis of the scaling violations $dF_2/d\ln Q^2 \sim \alpha_s$, or equivalently, from a QCD fit to the F_2 data, the strong coupling constant can in principle be determined. From an analysis of the 1993 data $\alpha_s(m_Z^2) = 0.120 \pm 0.005(\text{exp.}) \pm 0.009(\text{theor.})$ was obtained [61]. It is estimated [62] that ultimately, with an integrated HERA luminosity of 500 pb^{-1}, α_s can be extracted by a NLO QCD analysis from the HERA structure function data with an experimental error of $\Delta\alpha_s = 0.001 - 0.002$, and a theoretical uncertainty of ≈ 0.006. The theoretical errors could be reduced by higher than NLO calculations.

3.3 The Longitudinal Structure Function F_L

One can express the longitudinal structure function F_L in terms of the cross-section for the absorption of longitudinally polarized photons (see Sect. 2.1),

$$F_L = \frac{Q^2(1-x)}{4\pi^2\alpha} \cdot \sigma_L \approx \frac{Q^2}{4\pi^2\alpha} \cdot \sigma_L . \qquad (3.2)$$

Longitudinal photons have helicity 0 and can exist only virtually. For transverse photons with helicity ± 1 the spin is parallel to the direction of propagation, and the field vector perpendicular to it. In the QPM, helicity conservation at the electromagnetic vertex yields the Callan–Gross relation ($F_L = 0$) for scattering on quarks with spin 1/2. This does not hold when the quarks acquire transverse momenta from QCD radiation. Instead, QCD yields the Altarelli–Martinelli equation [63]

$$F_L(x,Q^2) = \frac{\alpha_s}{4\pi}x^2 \int_x^1 \frac{dz}{z^3}\left[\frac{16}{3}F_2(z,Q^2) + 8\sum_i e_{q_i}^2\left(1 - \frac{x}{z}\right)zg(z,Q^2)\right],$$

$$(3.3)$$

showing the dependence of F_L on the strong coupling and the gluon density. At small x, the second term with the gluon density is the dominant one. In fact $F_L(x,Q^2) = 0.3 \cdot [4\alpha_s/(3\pi)] \cdot xg(2.5x,Q^2)$ is not a bad approximation for $x < 10^{-3}$ [64].

The extraction of F_2 from the cross section measurement (2.1) so far had to make an assumption for F_L, because there existed no direct F_L measurements in the HERA regime. At large y, $y \approx 0.7$, this is a 10% correction. The argument can be turned around, and F_L can be extracted at large y from a measurement of the cross-section, assuming that F_2 follows

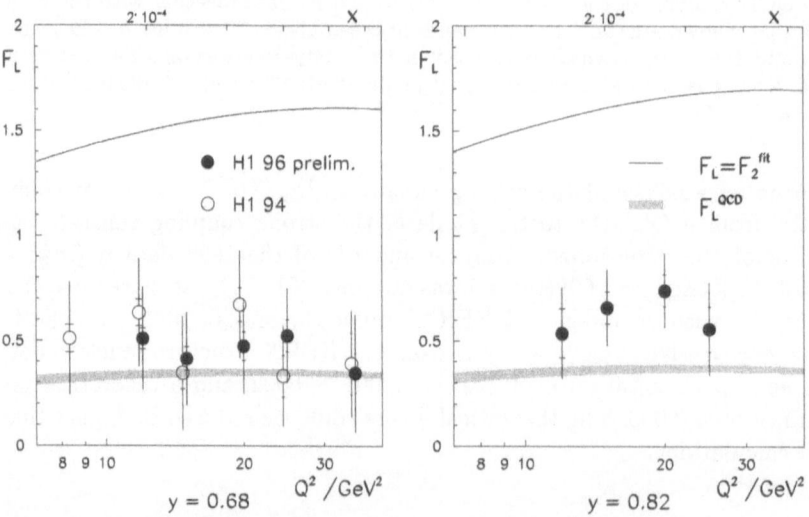

Fig. 3.4. The longitudinal structure functions F_L extracted from a QCD analysis of high y inclusive scattering data [42, 41] at two different y values. The shaded bands give the range of expectations for F_L from a QCD analysis of structure function data with $y < 0.35$. The full line represents the solution $F_L = F_2$

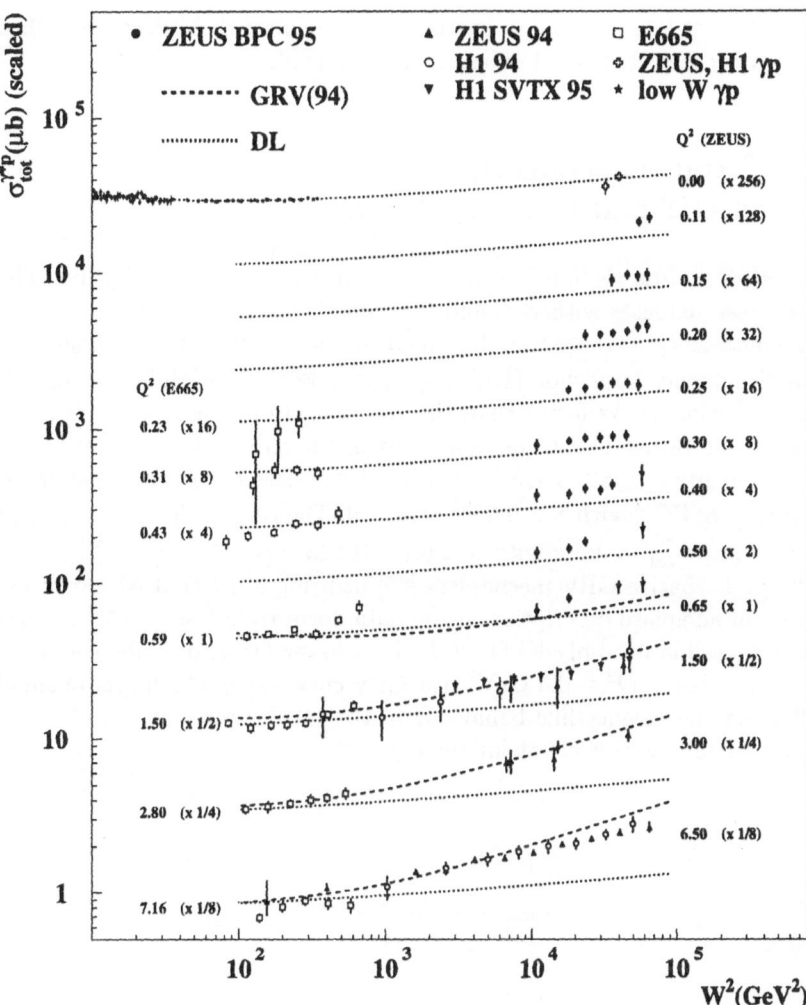

Fig. 3.5. The total γ^*p cross-section as a function of W^2 for different values of Q^2 [35]. The data for virtual photon–proton scattering [39, 40, 34, 35, 46] and photoproduction [66, 67, 68] are compared to the predictions by the soft Pomeron model of DL [69] and the perturbative QCD model by GRV [49]

a QCD evolution and can be extrapolated from measurements at smaller y. This procedure has been carried out by H1 [42, 41] (Fig. 3.4). The extracted F_L excludes the extreme possibilities $F_L = F_2$ and $F_L = 0$ and implies $R \equiv \sigma_L/\sigma_T = F_L/(F_2 - F_L) \approx 0.5$, since $F_2 \approx 1.5$. This is self-consistent with the gluon density extracted from the H1 QCD fit. It should be noted however that the F_L data points at high y are somewhat above the expectation from the QCD fit. More precise measurements are necessary to clarify the situation. A measurement of F_L without theoretical assumptions on the evolution of F_2 will be possible by measuring the DIS cross-section at HERA at two different centre of mass energies [65], because these data will allow one to vary y in (2.1) while keeping x and Q^2 fixed.

3.4 $\sigma_{tot}^{\gamma^*p}$ and the Transition Between DIS and Photoproduction

The total γ^*p cross section is shown as a function of W^2 in Fig. 3.5. The cross-section increases with W^2, and the slope increases with Q^2.

The data at $Q^2 = 0$ are well described by the soft Pomeron parametrization by Donnachie–Landshoff (DL) [30], $\sigma_{tot}^{\gamma^*p} \propto W^{2 \cdot 0.08}$. With increasing Q^2, the cross-section grows faster than the DL prediction. There appears to be a smooth but rather fast transition between the soft and the perturbative regions. Already at $Q^2 \approx 1$ GeV^2, a distinctively fast rise of the total cross-section $\sigma_{tot}^{\gamma^*p} \propto W^{2 \cdot \lambda}$ with $\lambda > 0.08$ is observed. The H1 fit result $F_2 \propto (1/x)^\lambda$ translates into $\sigma_{tot}^{\gamma^*p} \propto W^{2\lambda}$ with $\lambda \approx 0.2 - 0.4$ for $Q^2 > 1$ GeV^2.

The perturbative GRV mechanism [49] utilizing NLO DGLAP evolution provides an adequate description of the data down to $Q^2 \approx 1$ GeV^2. It has not been possible to apply NLO DGLAP at lower Q^2 to describe the data. For example, below $Q^2 \approx 0.4$ GeV^2 the GRV curves would turn over at small x, reflecting the valence-like behaviour of the GRV input partons at a low scale, by far failing to account for the data [35].

4. Models for Hadron Production in DIS

4.1 Reference Frames

In this section the different reference frames and the variables used to describe the hadronic final state are introduced. Rough expectations for the event properties are derived mainly from phase space arguments. They will be refined by perturbative QCD and specific hadronization models.

The Laboratory Frame

The HERA experiments are performed in the "laboratory frame", in which the detector is at rest, and the incoming electron and proton beams are collinear. The z-axis is defined by the proton beam direction. After the collision, the transverse momentum of the scattered electron is balanced by that of the hadronic system. The hadronic final state is however better studied in a frame where the transverse boost is removed, such that the virtual photon and the incoming proton are collinear. Photon and proton 4-vectors are denoted by $q = (E_{\gamma^*}, \mathbf{q})$ and $P = (E_p, \mathbf{P})$. With $E_p \gg m_p$ the proton mass can be neglected.

The Hadronic CMS

The hadronic centre of mass system (CMS) is defined by the condition $\mathbf{P} + \mathbf{q} = 0$. The invariant mass of the hadronic system is given by $W^2 = (P+q)^2$. The positive z-axis is usually defined by the direction of the virtual boson, q. All hadronic final state particles with 4-momentum $p = (E, p_x, p_y, p_z)$ which have $p_z > 0$ are said to belong to the current hemisphere, and all particles with $p_z < 0$ are assigned to the target or proton remnant hemisphere. Target and current systems are back to back and carry energy $W/2$ each. They are not necessarily collinear with the incoming proton. In the QPM however, a quark in the proton with 4-momentum xP absorbs the virtual photon and is scattered back with 4-momentum $xP + q$. The proton remnant retains the 4-momentum $(1-x)P$. With these assumptions, the scattered quark and the proton remnant are both collinear with the incoming virtual photon and the proton (Fig. 4.1). This is no longer the case if one considers intrinsic

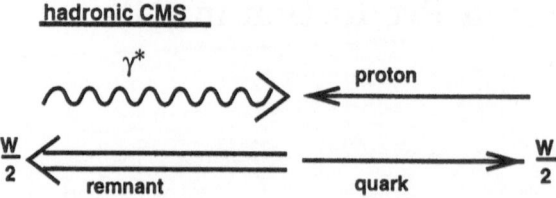

Fig. 4.1. The hadronic centre of mass system (CMS)

transverse momenta of the partons in the proton, or generates transverse momenta perturbatively by radiation.

The Breit Frame

The Breit frame (BF) is defined by the conditions that proton and virtual photon are collinear, and that the virtual photon does not transfer energy, just momentum. Since $q^2 = -Q^2$, $q = (0,0,0,Q)$ with the same orientation as in the CMS, and $P = (E_p, 0, 0, -E_p)$. In the QPM, the incoming quark absorbing the virtual photon does not change its energy, but reverses its longitudinal momentum with magnitude $Q/2$ (Fig. 4.2). The Breit system is also called the brick wall system, because in this picture the quark bounces back like a tennis ball off a brick wall. Neglecting the quark mass and transverse momenta, its 4-momentum is $xP = (Q/2, 0, 0, -Q/2)$ before and $(Q/2, 0, 0, Q/2)$ after the scattering. The incoming 3-momentum vector xP is merely reversed.

Again, all particles with $p_z > 0$ are assigned to the Breit current hemisphere. The CMS and BF are connected via a longitudinal boost. The division into current and target hemisphere is frame dependent. In both systems, the remnant diquark is a mere spectator in the sense that its momentum remains unchanged.

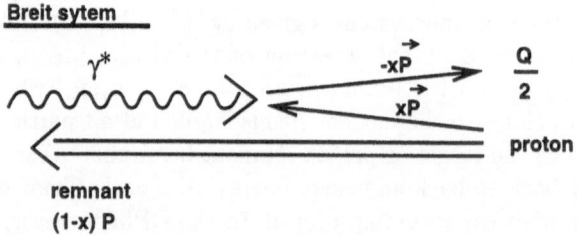

Fig. 4.2. The Breit frame

4.2 Kinematics and Variables

Let us consider a particle with 4-momentum $p = (E, \boldsymbol{p}) = (E, p_x, p_y, p_z)$ and mass m, which travels at an angle θ with respect to the z-axis. We define the transverse momentum $p_T = \sqrt{p_x^2 + p_y^2}$, the transverse mass $m_T = \sqrt{m^2 + p_T^2}$ and the transverse energy $E_T = E \cdot p_T / |\boldsymbol{p}|$.

The rapidity [70] of a particle is defined as

$$y := \frac{1}{2} \ln \frac{E + p_z}{E - p_z} = \ln \frac{E + p_z}{m_T} = \coth \frac{p_z}{E}. \tag{4.1}$$

The rapidity transforms under a boost in the z-direction with velocity β as

$$y \to y' = y - \coth \beta. \tag{4.2}$$

Hence the shape of the rapidity distribution is invariant against longitudinal boosts. A useful relation is

$$dy/dp_z = 1/E. \tag{4.3}$$

Often the mass of a measured particle is not known. One therefore defines the pseudorapidity

$$\eta := -\ln \tan \frac{\theta}{2}. \tag{4.4}$$

For $E \gg m$, $\eta \approx y$.

Energy and momentum of a particle are conveniently scaled by their kinematically allowed maximum, leading to the definitions

$$x_F := p_z/p_z^{\max} \qquad x_p := |\boldsymbol{p}|/p^{\max} \qquad x_E := E/E_{\max}. \tag{4.5}$$

For $E \gg m$ and $p_z \gg p_T$, these definitions converge: $x_F \approx x_p \approx x_E$. We shall use z as a generic symbol for the variables defined in (4.5).

The scaled longitudinal momentum x_F is known as Feynman-x. In the CMS, $x_F \approx 2 \cdot p_z/W$, with $-1 \leq x_F \leq 1$. Similarly, in the Breit frame current hemisphere $x_p = 2|\boldsymbol{p}|/Q$ with $0 \leq x_p \leq 1$. None of the variables defined in (4.5) is Lorentz invariant. A Lorentz invariant that is currently not being used at HERA is $z_h := (Pp)/(Pq)$. In the proton rest frame it gives the fraction of the energy transfer taken by the produced hadron, $z_h = E/\nu$. For large W and x_F, $z_h \gtrsim 0.1$ the differences between z and x_F are small.

The rapidity can take any values with $y_{\min} \leq y \leq y_{\max}$ (see Fig. 4.3), where

$$y_{\max} = \ln \frac{2p_z^{\max}}{m} \qquad y_{\min} = \ln \frac{2p_z^{\min}}{m}. \tag{4.6}$$

In the CMS, $-y_{\min} = y_{\max} \approx \ln W/m$. At HERA with $W \leq 300$ GeV, pions can be produced with $-7.7 \leq y \leq 7.7$. Neglecting transverse momenta, the CMS rapidity y is related to x_F (for $x_F > 0$) by

$$y_{\max} - y \approx \ln(1/x_F). \tag{4.7}$$

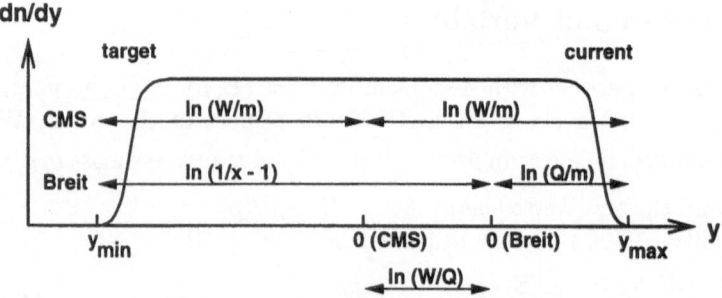

Fig. 4.3. The rapidity plateau. Indicated are the sizes of the target and current regions in the Breit and CM systems. The dimensions are chosen for typical HERA events with $W = 120$ GeV , $Q^2 = 30$ GeV^2 and for $m = m_\pi$

The difference of rapidities measured in the Breit and the CM systems is

$$y^{\text{BF}} - y^{\text{CMS}} = y_{\text{max}}^{\text{BF}} - y_{\text{max}}^{\text{CMS}} = \ln(Q/m) - \ln(W/m) = \ln(Q/W). \quad (4.8)$$

4.3 Simple Mechanisms for Hadron Production

Much can be learnt about the gross event features simply from phase space considerations and a few simple assumptions. We start with the simple model of independent fragmentation, applied to the scatterd quark and the remnant. In the following sections, this naive picture will be refined with perturbative QCD radiation and more sophisticated hadronization models.

The Lorentz invariant cross section for hadron production may be written as

$$E\frac{\mathrm{d}^3\sigma}{\mathrm{d}^3p} = \frac{\mathrm{d}^3\sigma}{\mathrm{d}\phi\mathrm{d}y p_T\mathrm{d}p_T}, \quad (4.9)$$

since $\mathrm{d}y/\mathrm{d}p_z = 1/E$. For a distribution that is isotropic in azimuth ϕ, the ϕ integration yields

$$\int_0^{2\pi} E\frac{\mathrm{d}^3\sigma}{\mathrm{d}^3p}\mathrm{d}\phi = 4\pi\frac{\mathrm{d}^2\sigma}{\mathrm{d}y\mathrm{d}p_T^2}. \quad (4.10)$$

We describe the inclusive densities to find a hadron h with energy fraction $z := E/E_q$ from the fragmentation of a quark q with fragmentation functions $D_q^h(z)$. ($D_q^h(z)$ is not a probability density, because its integral is not 1.) They scale approximately, that is they depend only on the fractional hadron energy z and not on the quark energy E_q [71]. The resulting hadron spectrum from quark fragmentation in the reaction $ep \rightarrow e'qX$ is then

$$\frac{1}{N}\frac{\mathrm{d}n_h}{\mathrm{d}z} = \frac{1}{\sigma_{\text{tot}}}\frac{\mathrm{d}\sigma}{\mathrm{d}z} = \frac{\sum_q e_q^2 f_q(x)D_q^h(z)}{\sum_q e_q^2 f_q(x)}. \quad (4.11)$$

Here σ_{tot} is the total event cross section, N the number of produced events, and n_h the number of produced hadrons of type h. $f_q(x)$ is the quark density function for flavour q in the proton.

In the simple model of independent quark fragmentation [72], a quark with energy E_q fragments into a hadron with energy E according to a distribution function $f(z)$, where $z = E/E_q$, see Fig. 4.4. The process is iterated with a quark carrying the remaining energy $(1 - z)E_q$ until there is no energy left to produce a hadron. It is assumed that in the fragmentation process only limited transverse momenta are produced, usually parameterized with a falling exponential distribution in p_T^2 with $\langle p_T^2 \rangle \approx 0.44$ GeV^2.

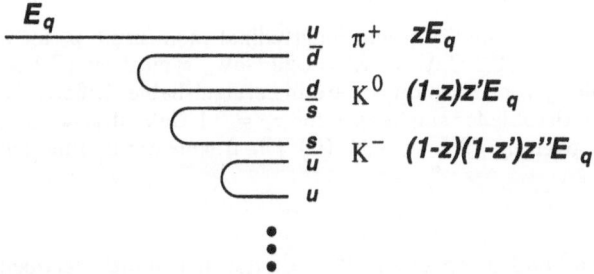

Fig. 4.4. Independent fragmentation à la Feynman–Field [72]. A u quark with initital energy E_q branches into a π^+ meson with energy $E = zE_q$, and a $\bar{u}$ quark with energy $(1 - z)E_q$, and so on

The model describes the gross features of fragmentation [72], namely energy independence, and for small z a behaviour

$$D_q^h(z) \propto 1/z. \tag{4.12}$$

Since $dy \approx z\,dz$, this gives a uniform number density of hadrons in rapidity, the so-called central (in the CMS) rapidity plateau:

$$\frac{1}{N}\frac{dn_h}{dy} \approx \frac{1}{N}\frac{dn_h}{dz} \cdot z \propto D_q^h(z) \cdot z = \text{const.} \tag{4.13}$$

Figure 4.5a gives the relation between the longitudinal momentum $p_z = z \cdot p_z^{\max} \approx z \cdot W/2$ for a typical HERA situation. $|x_F| < 0.1$ translates roughly to the rapidity region $|y| < 3$ (Fig. 4.5a).

A convenient parametrization for the fragmentation functions is

$$D_q^h(z) = a\frac{1}{z}(1 - z)^b. \tag{4.14}$$

Energy–momentum conservation requires

$$\sum_h \int_0^1 z D_q^h(z)\,dz = 1, \tag{4.15}$$

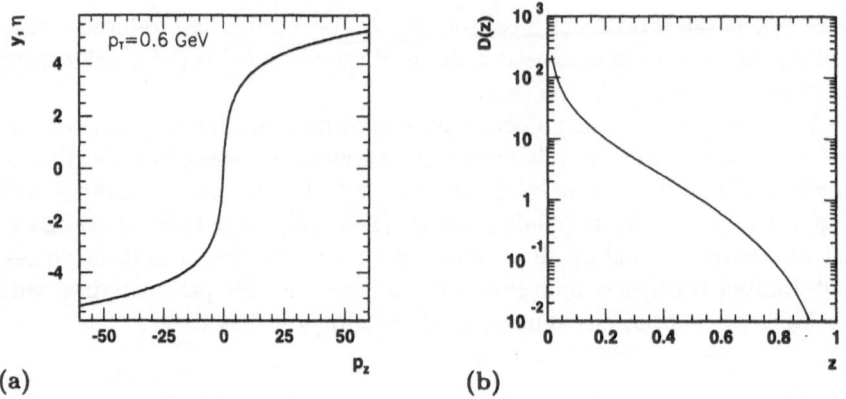

Fig. 4.5. (a) Relation between rapidity y and longitudinal momentum p_z for a particle with $m = m_\pi$ and $p_T = 0.6$ GeV at $W = 120$ GeV , typical for HERA. The curves for pseudorapidity η and rapidity y are indistinguishable. Differences between y and η become noticeable for smaller p_T; for $p_T = 0.1$ GeV it is at most 0.3 rapidity units. y_{max} is given by $p_z^{\text{max}} \approx W/2$. (b) The fragmentation function $D(z) = \frac{3.5}{z}(1-z)^{2.5}$

where the sum runs over all hadron species h. If we do not distinguish between hadron species, we can define $D_q(z) := \sum_h D_q^h(z)$. The normalization a is then fixed by

$$\int_0^1 zD_q(z)\mathrm{d}z = a/(b+1) = 1. \tag{4.16}$$

A useful parametrization for light quarks $q = u, d, s$, accurate to 20%, is [73]

$$D_q(z) = 3.5\frac{1}{z}(1-z)^{2.5}, \tag{4.17}$$

see Fig. 4.5b. The average number of hadrons per unit rapidity in the plateau region (z small) is then

$$\frac{1}{N}\frac{\mathrm{d}n}{\mathrm{d}y} \approx \frac{1}{N}\frac{\mathrm{d}n}{\mathrm{d}z}\cdot z = D_q(z)\cdot z \approx 3.5, \tag{4.18}$$

of which roughly 2/3 will be charged. The rapidity plateau is depicted in Fig. 4.3. There will be a kinematic fall-off for $y \to y_{\text{max}}$, y_{min} . The available rapidity range $y_{\text{max}} - y_{\text{min}}$ increases logarithmically with the quark energy (longitudinal phase space), and so does the average total hadron multiplicity.

The average total hadron multiplicity in a quark jet is given by

$$\langle n \rangle = \int_{z_{\text{min}}}^1 D_q(z)\mathrm{d}z, \tag{4.19}$$

where $z_{\text{min}} = m/E_q$ is determined by a typical hadron mass m. From

$$\mathrm{d}\langle n \rangle/\mathrm{d}z_{\text{min}} = -D_q(z_{\text{min}}) \approx -a/z_{\text{min}} \tag{4.20}$$

we obtain a logarithmic scaling law for the multiplicity

$$\langle n \rangle = a \ln \frac{E_q}{m} + c, \tag{4.21}$$

where c is the integration constant.

Though the independent fragmentation picture describes crudely the data on e.g. $e^+ e^- \to q\bar{q} \to$ hadrons, it has its limitations. It cannot be applied consistently for small z, $z \lesssim 0.1$, it is not Lorentz invariant (the constants a and c are frame dependent), and energy momentum conservation has to be enforced by hand after both quarks have fragmented. The lack of Lorentz invariance even leads to contradictions. Let W be the total CM energy. In the CMS, one calculates for the total multiplicity $\langle n \rangle = 2 \cdot (c + a \ln(E_q/m))$ with $E_q = W/2$. In a frame where one quark takes almost all of the energy, and the other one only very little, one gets a different value $\langle n \rangle = c + a \ln(E_q/m)$ with $E_q = W$

The discussion so far covered only the lowest order processes, where hadron production is determined by the assumption of limited transverse momenta and longitudinal phase space as a simple hadronization model. Though simple, it provides a good guideline for hadron production. QCD radiation modifys this simple picture and will be discussed next (Sect. 4.4). Afterwards improved hadronization models are introduced that do not suffer from the problems with the independent fragmentation model (Sect. 4.5).

4.4 Perturbative QCD Radiation

The perturbative part of QCD radiation is treated either with fixed order matrix elements, or with parton showers, or a combination of both. It depends on the application which treatment is most appropriate. In general, the matrix element will become more important with increasing hardness of the interaction. Perturbative QCD makes predictions for *partonic* final states, but *hadronic* final states are observed. To make contact with the experimentally accessible world, hadronization has to be taken into account. Models for the non-perturbative effects of hadronization are discussed in Sect. 4.5. Excellent reviews on perturbative QCD evolution and on hadronization are [74, 75]. Further information specific to DIS can be found in [76]. We start the discussion of perturbative QCD with matrix elements and parton showers. An alternative technique which is not based on Feynman diagrams is the dipole radiation approach, to be discussed later.

Matrix Elements

In fixed-order perturbation theory, the incoming parton flux is folded with the matrix element [77] for electron–parton scattering which leads to some final state parton configuration. In LO possible final state configurations for

eq scattering are (apart from the scattered electron) q (QPM), qg (QCDC), and for electron gluon scattering (BGF) $q\bar{q}$ (Fig. 4.6). In NLO, one additional parton can be emitted, and so on.

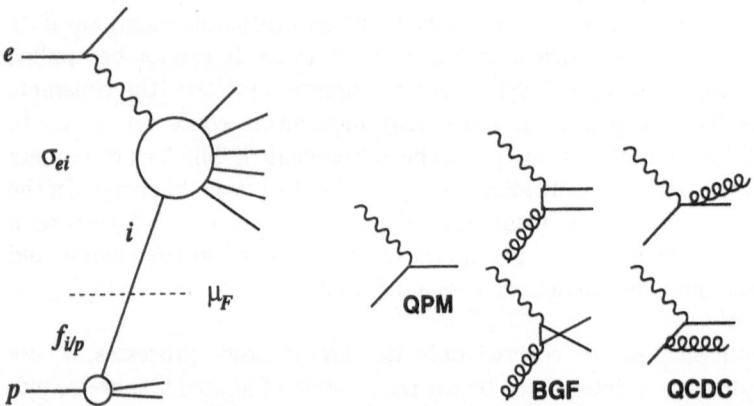

Fig. 4.6. Matrix elements in DIS. In leading order, scattering on a parton i can proceed via the graph that corresponds to the simple quark parton model (QPM), or the QCD analogue of Compton scattering (QCDC), or the boson–gluon fusion graph (BGF) in case the incoming parton i is a gluon. The graphs on the right hand are the LO realizations of the "blob" in the left graph

The calculations are ususaly done numerically. For state of the art NLO calculations the programs DISENT [78], MEPJET [79, 80], DISASTER++ [81] and JETVIP[82] are available. Originally such programs could only be used for jet analysis. The present programs are more flexible and allow also calculations of for example event shape variables.

Leading Log Parton Showers

A complete calculation to an order higher than NLO appears prohibitive. For many applications, where higher orders are important, the parton shower ansatz is used. For example, higher orders can be summed to all orders in the leading log approximation. In the DGLAP (leading log Q^2) approximation, they result in splitting functions which give parton emission probabilities.

When a parton with 4-momentum p radiates, it changes its virtuality $T^2 := -p^2$. The evolution parameter $t := \ln(T^2/\Lambda^2)$ increases in the initial state parton shower (space-like shower) by successive emissions, and decreases in the final state parton shower (time-like shower). The probability $\mathcal{P}$ that a branching $a \to bc$ will take place during a small change dt is given by the evolution equation [15]

$$\frac{d\mathcal{P}_{a\to bc}}{dt} = \int_0^1 dz \frac{\alpha_s(Q^2)}{2\pi} P_{a\to bc}(z). \tag{4.22}$$

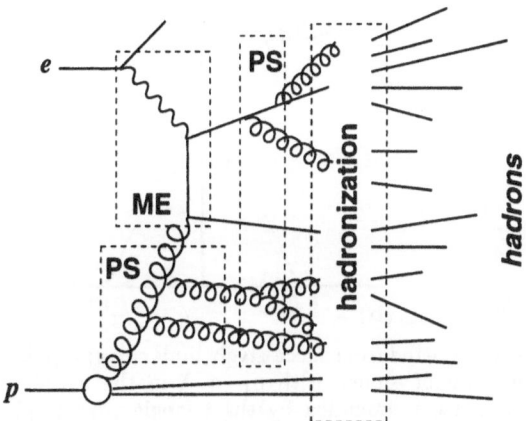

Fig. 4.7. Elements of an *ep* event generator. As an example an event is shown with the BGF matrix element (ME). The initial and final state parton showers (PS) produce additional emissions. Finally, the partonic final state (the parton level) is hadronized to yield the observable hadrons (the hadron level)

The functions $P_{a\to bc}(z)$ are just the Altarelli–Parisi splitting functions in (2.14), $P_{a\to bc} = P_{ba}$.

Such calculations are usually done with Monte Carlo event generators, where the parton shower evolution is simulated step by step according to the emission probabilites, until the whole event is generated. Perturbative evolution is stopped at some small scale T_0^2 of $\mathcal{O}(1\mathrm{GeV}^2)$, when it becomes unsafe to apply perturbation theory due to the growth of α_s.

When combined with the exact fixed order matrix element to take care of very hard emissions that are not properly covered with leading logs, the Monte Carlo event generators can be expected to provide a good representation of what is actually happening in DIS events[1] (Fig. 4.7). Ambiguities result from the way the matrix element and parton shower are combined ("matching"), coherence effects are treated, divergencies of the matrix element are cut-off, and from other approximations in the implementation. Unfortunately today's generators implement the matrix element and parton showers only to LO.

Examples of DIS Monte Carlo generators that are based on leading log DGLAP parton showers are LEPTO [83], HERWIG [84] and RAPGAP [85]. They are expected to be valid where the DGLAP approximation is valid. As DGLAP based models, the initial state parton shower is strongly ordered in k_T, increasing from the proton towards the matrix element.

The leading log approximation (LLA) can also be applied directly for specific observables, without utilizing an event generator. Most relevant is the modified leading log approximation (MLLA) which takes into account

---

[1] Remembering the double slit experiment, we should feel a bit uneasy about a statement of what is actually happening.

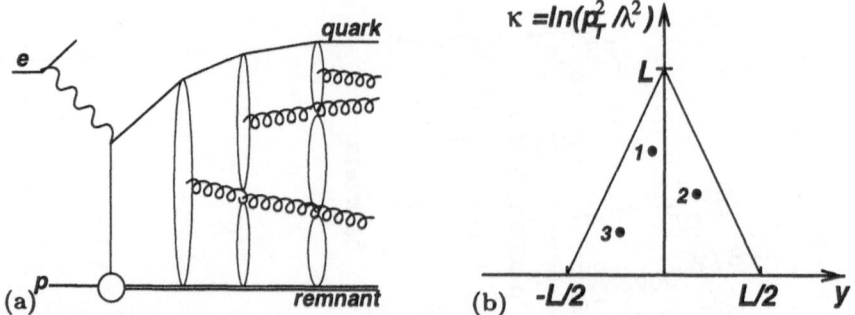

Fig. 4.8. (a) CDM in DIS: gluons are emitted from successively built colour dipoles. (b) The phase space for the emission of gluons with $p_T > \lambda$ in the variables $\kappa = \ln(p_T^2/\lambda^2)$ and y. The phase space is bounded by the triangle with height $L = \ln(W^2/\lambda^2)$. Subsequent emissions have decreasing p_T, $p_{Ti+1} < p_{Ti}$, but no ordering in y. If the emissions are reordered in y, they lose the p_T ordering

destructive interference between soft gluons, in conjunction with the assumption that the resulting parton spectra resemble the observable hadron spectra, up to a constant factor (local parton–hadron duality, LPHD).

The Colour Dipole Model

Another type of parton shower model is the colour dipole model (CDM) for QCD radiation [86]. The colour charges of the scattered quark and the remnant are assumed to form a colour dipole, from which gluons can be radiated (Fig. 4.8a). Subsequent gluon radiation emanates from dipoles spanned between the newly created colour charges and the others, and so on. To a good approximation it can be assumed that these dipoles radiate independently. The CDM uses the LO cross section for the emission of a gluon with transverse momentum p_T at rapidity y in the soft gluon approximation [87],

$$d\sigma = \frac{n_c \alpha_s}{2\pi} \frac{dp_T^2}{p_T^2} dy. \tag{4.23}$$

The cross section is uniform in rapidity and $\ln p_T^2$. Kinematically, the phase space is bounded by $|y| < \ln(W/p_T)$, where W is the total energy of the radiating system. Gluons with p_T above a certain cut-off, $p_T > \lambda$, lie inside the triangle in Fig. 4.8b.

In DIS one of the colour charges is not pointlike, the proton remnant. The suppression of radiation with short wavelength from an extended source is taken into account with a special parameter of $\mathcal{O}(1 \text{ fm})$ that describes the extendedness of the remnant.[2] The CDM is implemented in the program

---

[2] In the recent version also the colour charge of the scattered quark is spread out, depending on the "size" ($\propto 1/Q$) of the final state quark that can be resolved with the photon of virtuality Q^2.

ARIADNE [88]. In ARIADNE the QCDC graphs are covered by dipole radiation (and corrected to yield the exact LO matrix element contribution), but for the BGF graph the matrix element is used. Further radiation is then according to the CDM.

In contrast to the LL parton shower in the CDM it is not possible to distinguish between initial and final state radiation, or to reconstruct an "evolution path" for the parton in the proton that is hit by the virtual photon. Kinematically, the k_T phase space for radiation is bounded by the k_T of the previous emission. In rapidity the emission probability is uniform. Therefore the final gluon configuration is not ordered in k_T, if one arranges them according to increasing rapidity (see Fig. 4.8b). Rapidity is directly related to $\ln x_i$ for fixed W and k_{Ti} (see Sect. 8.1). In that respect the CDM is similar to what can be expected from BFKL evolution [89], though perhaps in a somewhat unconventional fashion [90]. Whereas the LL parton showers are connected with the intuitive picture of an evolution path, it is probably fair to say that the significance of the dipole model is not yet understood very well. We shall see that this model provides effortlessly overall the best description of most final state data. This is probably the main reason why it is being discussed, whether or not it has a deeper physical justification not yet appreciated by the community.

The Linked Dipole Chain Model – LDC

The CCFM approach for the hadronic final state has been reformulated in a picture similar to the CDM, giving rise to the linked dipole chain model (LDC) [24]. This model is particularly interesting, as it should converge to the DGLAP and BFKL predictions in their respective regions of validity, and handle properly cancellations between real and virtual emissions for the final state. In the LDC also the case is treated where the largest p_T of the process is not attached to the virtual photon vertex. Some promising results from a Monte Carlo implementation of the LDC have been obtained [91], but the program is not yet publicly available.

Coherence

Quantum mechanical interference is an important issue for the hadronic final state. With matrix elements that sum and square amplitudes interference effects are taken into account naturally. Coherence effects in parton showers that rely on emission probabilities, not amplitudes, require special attention. One may distinguish (a) interference between initial and final state radiation; and (b) interference between partons emitted either in the initial state or in the final state parton shower.

In the CDM there is no distinction between initial and final state radiation, so in the dipole approximation problem (a) does not arise. Also soft gluon interference is automatically taken into account by the dipole mechanism. In other generators such interference effects can be realized by phase space restrictions for parton emission. For example, the effect of destructive interference between subsequent emissions in a parton shower can be approximated by imposing a posteriori "angular ordering": only emissions are allowed with a smaller opening angle than the previous one, $\theta_{i+1} < \theta_i$.

The physical reason for this condition is that large wavelength quanta cannot be emitted from dipole sources with small transverse dimensions. This can be seen easily for e^+e^- cascades [92]. Consider the example Fig. 4.9a, where a virtual photon branches into an e^+e^- pair, and the e^- radiates a photon. The formation time of the final $ee\gamma$ state can be estimated as the lifetime of the intermediate e^- from its off-shellness M, $\tau = \gamma \cdot \tau_0 = E'/M \cdot 1/M$. We have $M^2 = (p+k)^2 \approx 2EE_\gamma(1-\cos\theta_2)$, and $1-\cos\theta_2 \approx \theta_2^2/2$ for small angles. Since for soft photons, $E_\gamma \ll E' \approx E$, $\tau \approx 1/(E_\gamma\theta_2^2)$. During this time the e^+e^- pair have separated a transverse distance $\delta r \approx \tau\theta_1$. The transverse wavelength of the emitted photon is $\lambda_T = 1/k_T$ with $k_T = E_\gamma\theta_2$. If λ_T is larger than δr, the photon cannot resolve the individual charge of the e^-, it rather sees the combined charge of the e^+e^- pair, which is 0. For allowed emissions it follows $\delta r > \lambda_T$, or $\theta_1 > \theta_2$, angular ordering.

In QCD the situation is a bit more complicated, because the incoming parton is colour charged (Fig. 4.9b). Parton b cannot be radiated incoherently from parton v when $\theta_2 > \theta_1$. In that case it would see the combined colour charge of a and v, which is the colour charge of the incoming parton i. Parton b with $\theta_2 > \theta_1$ can therefore be treated as effectively being radiated from i before a branches off, thus restoring angular ordering.

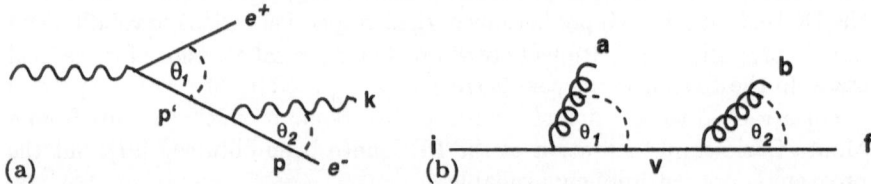

(a)   (b)

Fig. 4.9. (a) Angular ordering in an electromagnetic cascade. The 4-momenta of the outgoing and intermediate electron are $p = (E, \boldsymbol{p})$, $p' = (E', \boldsymbol{p}')$, and of the outgoing photon $k = (E_\gamma, \boldsymbol{k})$. (b) Angular ordering in a QCD cascade

4.5 Hadronization

Experimentalists follow two different philosophies on how to prepare their hadronic final state data.

- In one approach, the data are corrected just for detector effects in order to remove any experimental bias and be able to compare the data to other experiments. The data are corrected "to the hadron level"; they represent a measurement of observable hadrons. The data can also be compared to QCD predictions that include hadronization effects, but not directly to perturbative QCD calculations (unless one is interested in studying the difference between observable hadrons and partonic calculations).

- In the other approach, a hadronization model is used to correct the data for hadronization effects back "to the parton level". The data can then be compared directly to perturbative QCD calculations. A problem of principle arises: partons are not observable, so strictly speaking an observable "at the parton level" does not exist. With great care one may proceed nevertheless. One has to specify the parton level in perturbative QCD (like NLO, LLA, including cut-offs, etc.), and correct the data for all other effects beyond that: higher order corrections plus hadronization effects. The observable thus becomes scheme dependent. There is another price to pay: the data have picked up a theoretical bias due to the assumptions on higher order and hadronization effects. In practice this bias is studied by comparing different Monte Carlo generators. The encountered differences are treated as systematic errors of the measurement.

We summarize the main approaches to hadronization, some of which have already been mentioned. It has to be emphasized that the boundary between perturbative QCD evolution and hadronization is not uniquely defined. Ideally, predictions for final state observables would not depend on how the boundary is defined, for example on the lower cut-off for virtualities in the parton shower.

We start with hadronization models that are implemented in Monte Carlo generators (independent fragmentation, string and cluster model). Other approaches that model only specific aspects of hadron production are discussed later.

Independent Fragmentation [72]

From the fragmenting parton mesons are produced, carrying a certain fraction z of the original energy, see Sect. 4.3 and Fig. 4.4. The branching is repeated with the remaining energy until all the energy is used up. In the physical picture for independent quark fragmentation the quark forms a meson with an antiquark from a $q\bar{q}$ pair, with the remaining quark continuing fragmentation. It is assumed that the distribution of z is energy independent, leading to scaling fragmentation functions. Furthermore, all partons of an event fragment independently. That leads to inconsistencies like non-conservation of energy–momentum, that have to be cured by hand.

Transverse momentum components are assumed to be Gaussian distributed[3] [94], which results in an exponentially falling distribution in p_T^2.

---

[3] The transverse momentum p_T which a primary hadron receives from the fragmentation process can be described by Gaussians in p_x and p_y with widths σ:

From the parametrizations used in fragmentation models [94, 93], $\langle p_T^{\text{frag}\,2}\rangle \approx$ (0.3 GeV)^2 is obtained for *primary* hadrons.

String Fragmentation [95]

In the Lund string model, a quark and an antiquark moving apart stretch a colour field between them. The field is thought to be string-like with constant energy density per unit length of $\mathcal{O}(1$ GeV /fm), and with transverse dimension of $\mathcal{O}(1$ fm). When the stored energy becomes large enough, the string can break up and create a $q\bar{q}$ pair from the vacuum (Fig. 4.10a). The newly created q and $\bar{q}$ terminate the loose string ends, such that the new string pieces are themselves colour-neutral. The process is iterated with the new strings until all the available energy is used up. Transverse momentum components result from a tunnelling process; they are parametrized as in (4.24). The string model gives rise to scaling fragmentation functions, a rapidity plateau with uniform particle density, and a logarithmic increase of particle multiplicity $\langle n \rangle = c + a \ln W$, where W is the string invariant mass.

String fragmentation can be formulated in a covariant way; it does not suffer from the above-mentioned shortcomings of independent fragmentation. Furthermore, some quantum mechanical interference effects in gluon radiation are taken into account with string fragmentation, due to the way gluons are treated. Carrying two colour charges, gluons are always the endpoints of two strings. In a $q\bar{q}g$ configuration the gluon is realized as a kink in the colour connection between the $q\bar{q}$ pair. Destructive gluon interference leads to a suppression of radiation (and finally hadron production) in the region opposite to the gluon. Because in the string model there is no string spanned directly between the q and the $\bar{q}$ this "string effect" [96] appears naturally in the Lund model.

$$\frac{\mathrm{d}^2 n}{\mathrm{d}p_x \mathrm{d}p_y} = \frac{1}{\sigma\sqrt{2\pi}} \exp\left(\frac{-p_x^2}{2\sigma^2}\right) \frac{1}{\sigma\sqrt{2\pi}} \exp\left(\frac{-p_y^2}{2\sigma^2}\right). \tag{4.24}$$

This leads to an exponentially falling distribution in p_T^2 with $\langle p_T^2 \rangle = 2\sigma^2$,

$$\frac{\mathrm{d}n}{\mathrm{d}p_T^2} = \frac{1}{2\sigma^2} \exp\left(\frac{-p_T^2}{2\sigma^2}\right). \tag{4.25}$$

It follows that the distribution in p_T is neither Gaussian nor exponential:

$$\frac{\mathrm{d}n}{\mathrm{d}p_T} = \frac{p_T}{\sigma^2} \exp\left(\frac{-p_T^2}{2\sigma^2}\right), \tag{4.26}$$

with $\langle p_T \rangle = \sqrt{\frac{\pi}{2}}\sigma = \frac{\sqrt{\pi}}{2}\sqrt{\langle p_T^2 \rangle}$. The default parameter in the Monte Carlo program JETSET [93] is $\sigma = 0.36$ GeV.

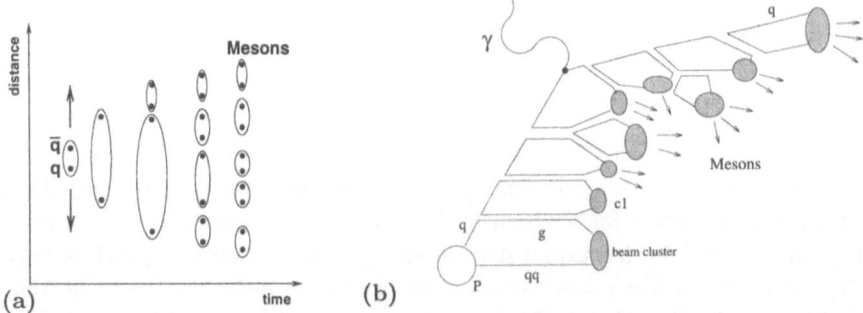

Fig. 4.10. (a) String fragmentation. $q\bar{q}$ pairs are created when the stretched string between the separating $q\bar{q}$ pair breaks up. Mesons are formed when the string energy is too small to create further $q\bar{q}$ pairs. (b) Cluster fragmentation. Gluons are represented by double lines, corresponding to the colour and anti-colour they carry. After the perturbative evolution gluons are split non-perturbatively into $q\bar{q}$ pairs, from which colour-neutral clusters are formed. They decay into mesons

Cluster Fragmentation [97]

According to the idea of "preconfinement" [98], colour-connected partons tend to be close in phase space towards the end of the perturbative phase. This property is exploited in the cluster fragmentation model. After the perturbative evolution all gluons are forced to split into $q\bar{q}$ pairs. Colour connected $q\bar{q}$ pairs are combined into low-mass colour-neutral clusters, which decay into mesons according to phase space (Fig. 4.10). Due to the small cluster mass, relatively few parameters are needed to describe the cluster decay. Nevertheless, large mass clusters may occur; they are split longitudinally in a fashion similar to the string model before they are allowed to decay.

Local Parton–Hadron Duality (LPHD)

The hypothesis of local parton–hadron duality (LPHD) [99] is made to allow predictions for hadron spectra from perturbatively calculated parton spectra. It states that any hadron cross-section depending on a quantity ζ is related to the corresponding partonic cross-section simply by a constant factor, $d\sigma_h/d\zeta_h = c \cdot d\sigma_p/d\zeta_p$. The idea behind this seemingly surprising ansatz is that if perturbative evolution is used for multi-parton emissions down to a very low scale, there is not much energy left for hadronization to change the shape of the distribution. This hypothesis could be applied successfully to a variety of phenomena, but one should keep its limitations in mind [100]. For example, when a quark and antiquark separate and do not radiate any gluons (unlikely, but possible), LPHD would predict a gap in rapidity in between, devoid of any hadrons. A more physical assumption is that the colour field stretched between quark and antiquark leads to the production of hadrons filling the gap. At HERA, the LPHD hypothesis has been applied to charged particle spectra, see Sects. 5.2 and 6.1.

Fragmentation Functions

The number density to find a hadron h with energy fraction $z := E/E_q$ from the fragmentation of a quark q is described by a fragmentation function $D_q^h(z)$, and similarly for a gluon g. They scale approximately, that is they depend only on the fractional hadron energy and not on the quark energy [71]. QCD effects like gluon radiation off the quark make the fragmentation functions scale-dependent: $D_q^h(z) \to D_q^h(z,t)$, where the scale t is provided by the hard process (Fig. 4.11). While the fragmentation functions themselves are not calculable in perturbative QCD, their scale dependence can be calculated in the same way as the scaling violations of the structure functions in DIS [101].

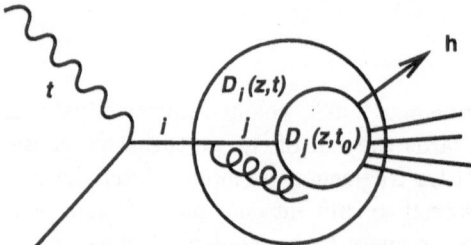

Fig. 4.11. Scaling violations of fragmentation functions. In this case the scale is given by the virtuality of the photon, $t = Q^2$

The fragmentation functions obey an evolution equation

$$\frac{\partial D_{h/i}(z,\mu_D^2)}{\partial \ln \mu_D^2} = \sum_j \frac{\alpha_s(\mu_D^2)}{2\pi} \int_z^1 \frac{du}{u} P_{ji}(u,\alpha_s(\mu_D^2)) D_{h/j}\left(\frac{z}{u},\mu_D^2\right). \quad (4.27)$$

$t = \mu_D^2$ is the factorization scale between the parton production process and the parton fragmentation process. The splitting functions P_{ji} for branchings $i \to j$ are in leading order the same as for the parton densities in DIS (2.14), but differ in higher order [101]. They have been calculated up to next-to-leading order [102].

The fragmentation functions are universal, that is process-independent. For example, quark fragmentation functions measured in $e^+e^- \to$ hadrons can be used to predict the hadron distribution in the quark jet for $ep \to e'qX$.

High Energy and Power Corrections

Due to the limited p_T generated in hadronization, it appears plausible that with increasing energy Q the relative importance of hadronization effects diminishes for suitably chosen observables. It is expected that for infrared safe quantities (see Sect. 2.6) like jet observables and some event shape variables their perturbative prediction approaches the value that is observable with hadrons. The difference, due to higher orders and hadronization, should decrease like an inverse power of the energy $\propto (1/Q)^a$, a "power correction" [103]. Power corrections have been investigated for thrust and other event shape variables at HERA, see Sect. 6.2.

4.6 Rapidity Gaps

At HERA there exists a class of events known as "rapidity gap events" [104, 105, 106, 107]. They have given rise to tremendous activity, which deserves a review in its own right (recent summaries are [3]). Here as much is covered as is relevant for our purpose.

Rapidity gap events exhibit a sizeable region of rapidity without any hadronic activity – the rapidity gap, see Fig. 4.12. In normal events a colour

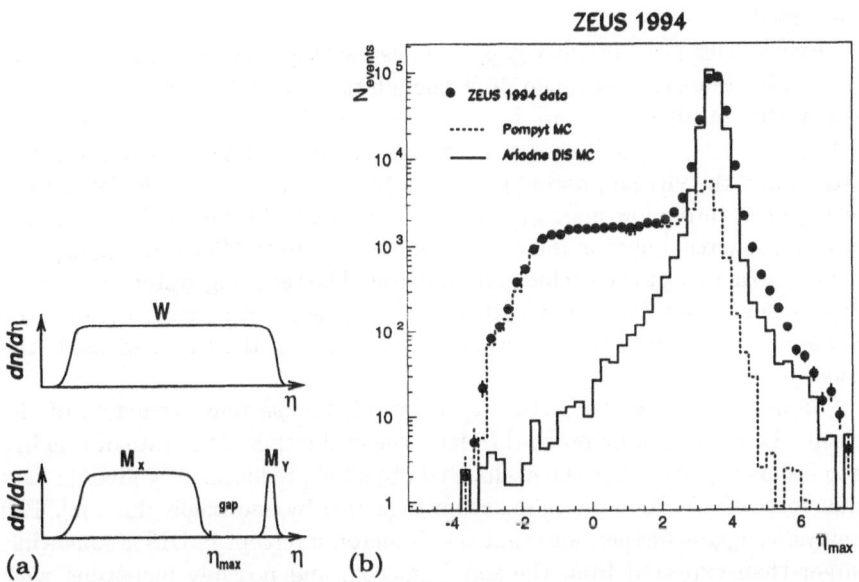

Fig. 4.12. (a) Rapidity distribution for normal events (*top*) and for rapidity gap events (*bottom*). M_X and M_Y are the invariant masses of the two systems X and Y separated by the largest gap in the event. The system Y is the proton or its remnant. The proton direction is to the right. η_{max} is the forward rapidity edge of the system X. A large gap implies a small mass M_X. (b) η_{max} distribution from ZEUS (uncorrected) in the laboratory system. The data are compared to a normal DIS model (ARIADNE [88]) and to a model for rapidity gap events based on Pomeron exchange (POMPYT [108]). For most events $\eta_{max} \approx 3.5$, given by the forward edge of the main calorimeter. Rapidity gap events are found in the tail at small η_{max}

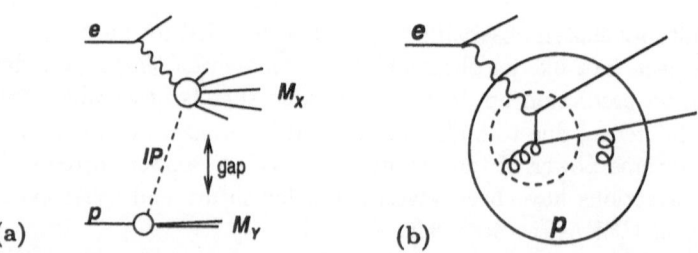

Fig. 4.13. (a) DIS with Pomeron exchange. The proton momentum fraction carried by the Pomeron, x_{IP}, can be reconstructed by $x_{IP} = (M_X^2 + Q^2)x/Q^2$. The variable $\beta := x/x_{IP} = Q^2/(M_X^2 + Q^2)$ gives the Pomeron momentum fraction carried by the parton that is hit by the virtual photon. (b) DIS with colour rotation. After the hard interaction (BGF, *broken circle*) a soft gluon is interchanged with the proton

field forms between the scattered quark and the proton remnant, leading to a homogeneous distribution of hadrons populating the available rapidity range. Large gap sizes between neighbouring hadrons from normal fragmentation are expected to be unlikely (though possible). Large gaps can be produced however in processes where two systems are produced that are not colour connected.

Events with large rapidity gaps had been anticipated for HERA from processes with Pomeron exchange [109], though the large rate came as a surprise: about 10% of all DIS events have a large rapidity gap.[4] In the Ingelman–Schlein model [110] a Pomeron is exchanged in the t channel between the proton and the virtual photon, see Fig. 4.13a. The proton may be left intact, or fragment into a low-mass system $M_Y \lesssim 1.6$ GeV. In the Ingelman–Schlein model, the virtual photon interacts with a parton in the Pomeron, thus probing the structure of the exchanged Pomeron. The resulting system with mass M_X is not colour connected with the proton remnant system, because the Pomeron carries no colour. A rapidity gap between the two systems is the consequence.

Supposing the above picture to be valid, the *partonic structure* of the elusive Pomeron can be probed! In fact, the evaluation of the rate of rapidity gap events suggests that the exchanged object is predominantly gluonic, and that most of the Pomeron momentum is carried by one single gluon [3]. The analyses suggest furthermore that the Pomeron intercept in DIS is somewhat larger than expected from the soft Pomeron, and possibly increasing with Q^2 [3].

Other models assume that rapidity gaps are due to changes in the colour structure of the event *after* the hard interaction has taken place [111, 112]. At low x, most events are due to photon–gluon fusion, resulting in a colour

---

[4] Often these events are called "diffractive", or being due to a diffractive process. The meaning of these terms is not yet uniquely defined amongst the experiments and theorists. We therefore prefer the term "large rapidity gap", because the existence of a gap of a certain size is well defined event by event.

octet–antioctet configuration in the final state. In the string model, there
would be two fragmenting strings spanned (Fig. 4.14a). It is assumed that
while the produced quark and antiquark travel to leave the proton, soft gluons are exchanged between them and the proton remnant, see Fig. 4.13b.
These gluons do not change the momenta significantly, but may rotate the
colour configuration of the quark–antiquark pair. The final configuration may
turn out to be a colour singlet, in which case there would be no colour field
connecting it with the remnant (Fig. 4.14c). The attractive feature of this
model is an explanation of the absolute rate of rapidity gap events. If soft
gluon exchange results in a random colour configuration, the ratio of colour
singlet to colour octet configurations would be 1:8, and $1/9 \approx 10\%$! In a related semiclassical approach, partonic fluctuations of the virtual photon are
scattered by the colour field of the proton [113].

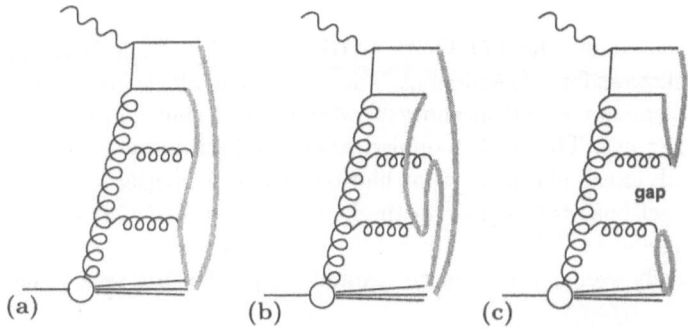

Fig. 4.14. String configurations for the normal case (**a**), and after soft colour
interactions, leading to either longer strings (**b**) or a rapidity gap (**c**)

Both models describe at least qualitatively the HERA data on rapidity
gap events. Detailed studies will be necessary to discriminate between them
(for example in the final state of the system X [114, 115]), or perhaps discover
them to be not exclusive at all.[5] For the discussion of the hadronic final state
in DIS it will be important to keep the existence of rapidity gap events and
possible production mechanisms for them in mind. Some analyses [116, 117,
118, 119, 120, 121] explicitly exclude rapidity gap events, others don't, and
some study the difference between the two samples. The possible existence
of a mechanism to change the colour configuration will have a bearing on
the interpretation of certain final state data at small x (Sect. 8.1). Colour
reconnections are also being discussed in connection with WW fragmentation
at LEP [122, 123].

---

[5] Pomeron exchange where most of the momentum is carried by one gluon in
the Pomeron is not so much different from boson–gluon fusion, where the hard
dynamics is determined by the incoming gluon, and soft gluons are only needed
for colour neutralization.

4.7 Monte Carlo Generators

We briefly summarize the DIS Monte Carlo generators. They incorporate the QCD evolution in different approximations and utilize phenomenological models for the non-perturbative hadronization phase. The events are generated according to the electroweak cross section $\mathrm{d}^2\sigma/\mathrm{d}x\mathrm{d}Q^2$ with experimentally determined input parton densities. QED radiative effects are treated with special programs [124, 125].

The Monte Carlo results depend on the model parameters, which can be adjusted to the data. Information on generator tuning can be found in [126, 127, 128, 76]. In general, the generators now give an overall satisfactory description of the data. None of the generators however is able to describe all aspects of the data.

LEPTO

LEPTO [83] is based on the LO QCD matrix element with leading log DGLAP parton showers for soft emissions. Therefore the model is often called MEPS (Matrix Elements + Parton Showers). Angular ordering is imposed to model colour coherence. The rather sophisticated Lund string model [95] as implemented in JETSET [93] is used for hadronization. A special feature is "soft colour interactions" (SCI) [111] in the hadronization phase. They lead to string rearrangements. This feature is now default, because it describes at least qualitatively events with a large "rapidity gap" that are seen in the data.

RAPGAP

Originally, RAPGAP [85] was created for rapidity gap events. They are described by scattering on a colour neutral object, the Pomeron, which is emitted from the proton according to a flux factor, and whose internal structure is described by parton density functions. The present version includes also normal DIS event generation, very similar to LEPTO. In the latest version it is also possible to simulate a resolved (not pointlike) component of the virtual photon.

HERWIG

HERWIG [84] is based on LL parton showers, with additional corrections to describe properly the hard interaction. Final state radiation is angular ordered, initial state radiation is ordered in $E \cdot \theta$, where E and θ are the energy and angle of the radiated parton. The HERWIG philosophy is to model the perturbative phase as accurately as possible, and to hadronize with a relatively simple cluster fragmentation model [97]. Though there is no mechanism foreseen for rapidity gap events, it turns out that they are generated at a surprisingly large rate. Also with HERWIG, a resolved component of the virtual photon can be simulated.

ARIADNE

In ARIADNE [88] perturbative QCD radiation is modelled with radiating colour dipoles according to the CDM. The BGF graph is added by hand with its LO matrix element. Hadronization is performed with the string model as implemented in JETSET. ARIADNE allows one to model rapidity gaps either by scattering on a Pomeron, or by colour reconnections, but these options are by default not activated.[6]

---

[6] It was found that colour reconnections in combination with the colour dipole model do not give a good description of rapidity gap events at HERA [129].

5. General Event Properties

5.1 Energy Flow

The first analysis of the hadronic final state in DIS at HERA was based on 88 events recorded in 1992 [130]. The measured energy flows and particle spectra were in rough agreement with what was expected from some models including QCD radiation and fragmentation, and exluded certain other radiation scenarios.

The flow of transverse energy E_T as a function of pseudorapidity η provides a very simple, global characterization of the hadronic final state. It is measured with the calorimeters and includes all produced particles, except the scattered electron. The E_T flow as a function of pseudorapidity is defined as

$$\frac{1}{N}\frac{\mathrm{d}E_T}{\mathrm{d}\eta} = \frac{1}{\sigma}\sum_h \int E_T \frac{\mathrm{d}^2\sigma_h}{\mathrm{d}\eta\mathrm{d}E_T}\mathrm{d}E_T, \tag{5.1}$$

where the sum extends over all particle species h with cross sections σ_h. The definition is analogous for the E_T flow as a function of azimuth ϕ. The distribution is normalized to the total number of events N, corresponding to the total event cross section σ.

It is instructive to examine the flow of hadronic transverse energy as seen in the HERA detectors in the laboratory frame, Fig. 5.1. In this frame one measures according to (5.1) the energy flow transverse to the beam line as a function of pseudorapidity $\eta = -\ln\tan(\theta/2)$ and azimuth ϕ. Here θ is the angle with respect to the proton beam axis (laboratory $+z$), and ϕ is the angle between the energy deposition and the scattered electron in the plane transverse to the beam line. One sees the current jet from the scattered quark emerging opposite to the scattered electron in ϕ (Fig. 5.1 right). The rapidity region from the current jet towards the remnant is filled with particles due to the colour field stretched between the scattered quark and the remnant (the main calorimeter acceptance ends at typically $\eta = 3.5$), see Fig. 5.1 left.

Significantly more transverse energy is observed between the current jet and the beam pipe than is expected from the QPM without QCD radiation, but including fragmentation. The QPM predicts an E_T of roughly 1 GeV per unit pseudorapidity; one arrives at a similar number from the simple considerations in Sect. 4.3, assuming 3.5 hadrons per unit rapidity with an average

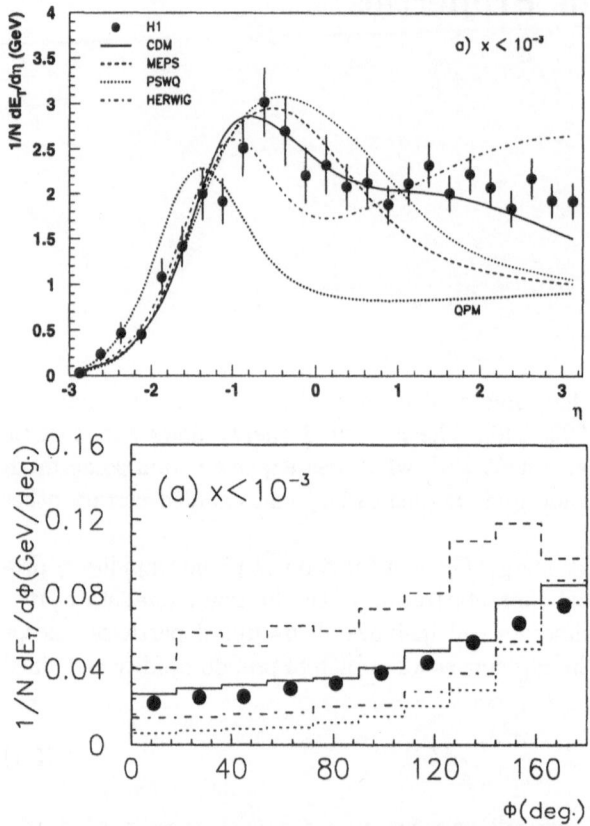

Fig. 5.1. Transverse energy flow in the laboratory frame for $x < 10^{-3}$. The plots are normalized to the number of observed events N. *(Top)* E_T as a function of pseudorapidity η by H1 [116]. The proton beam direction is to the right. Shown are statistical and systematic errors added in quadrature, except for a 6% scale error. The data are compared to the QPM without QCD radiation, and to models including QCD radiation: CDM (colour dipole model, ARIADNE 4.03), MEPS (matrix element plus parton showers, LEPTO 6.1), PSWQ (parton shower only, LEPTO 6.1, with maximal virtuality in the parton shower $W \cdot Q$), and HERWIG (HERWIG 5.7). *(Bottom)* E_T as a function of azimuthal angle ϕ with respect to the electron scattering angle in the transverse plane for clusters with $\theta > 10°$ by ZEUS [131]. Comparisons are made to different model predictions: matrix element only (ME, *dash-dotted line*), matrix element plus parton shower (MEPS, *full line*), parton showers only (PS) with either W^2 (*dashed line*) of Q^2 (*dotted line*) as maximum for the virtuality in the PS. The H1 data are corrected for detector effects. The ZEUS data are uncorrected, and compared to the models including detector simulation

p_T of 0.35 GeV. The extra E_T can be attributed to hard and soft gluon radiation. The models for these processes varied a lot in their predictions, but have evolved since.

In Fig. 5.2 the latest preliminary H1 data on energy flows in the hadronic CMS [118] are compared to the model LEPTO 6.4 with different treatments for perturbative QCD radiation. Without QCD radiation, a flat rapidity plateau with $E_T \approx 0.7$ GeV per unit rapidity is expected. QCD radiation as expected from the hard matrix element generates E_T in the current hemisphere only. It is shown below that the E_T in the current hemisphere depends on Q^2. With parton showers only, a reasonable description of the data is achieved. For this observable, the parton shower already covers most of the relevant phase space, including the matrix element. The best description of the data is obtained by combining radiation according to the LO matrix element with parton showers.

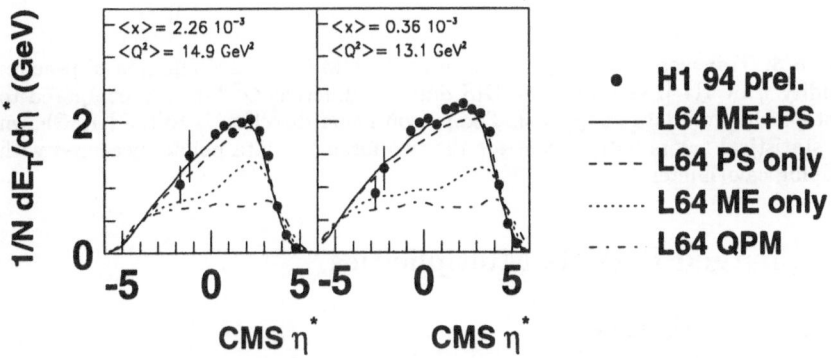

Fig. 5.2. The E_T flow from H1 [118] at $\langle x \rangle \approx 0.002$ (*left*) and 0.0004 (*right*) for $\langle Q^2 \rangle \approx 14$GeV^2. The data are in the hadronic CMS (proton direction to the left). Shown are statistical errors only. The data are compared to LEPTO 6.4 with different treatments of perturbative QCD effects: QPM: no radiation; ME: matrix element only; PS: parton showers only; ME+PS: matrix element plus parton shower (default)

The energy flow has also been studied as a function of Q^2 (Fig. 5.3). At low Q^2 the rapidity plateau is almost flat. With increasing Q^2, the E_T in the "photon fragmentation region" (loosely defined as far in the current hemisphere) increases considerably. In that region, the virtuality of the photon governing hard interactions plays an important rôle. At central rapidity, the Q^2 dependence is comparatively small. This supports qualitatively the picture that particle production at central rapidity is independent of the type of colliding particles, be they π, K, p or a virtual photon γ^* [132].

In Sect. 8.2 the E_T flow is discussed in more detail in connection with low x physics. Further data on hadronic energy production are available from H1, namely energy–energy correlations [116, 118], and the distribution of transverse energy at central rapdity [118].

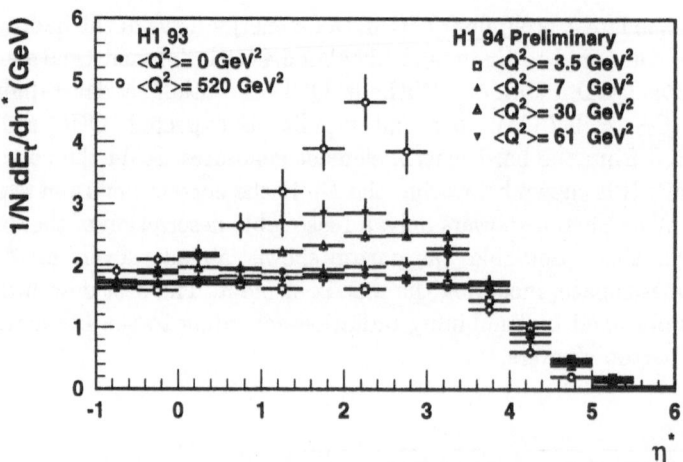

Fig. 5.3. Transverse energy production in the CMS. E_T as a function of pseudo-rapidity η by H1 [133, 118]. The DIS data for different Q^2 bins are compared to photoproduction data at $Q^2 \approx 0$. The proton beam direction is to the left. Shown are statistical errors only, except for the two foremost data points measured with the plug calorimeter

5.2 Charged Particle Multiplicities

Average Multiplicity

The multiplicity of charged particles in the CMS current hemisphere ($x_F > 0$) has been measured by H1 [119] with their central tracking chamber for events[1] at $\langle Q^2 \rangle \approx 23$ GeV^2 in the kinematic range 80 GeV $< W < 220$ GeV (Fig. 5.4a), thus reaching fragmentation energies not yet accessible at LEP. From phase space arguments alone, one would expect $\langle n \rangle \propto \ln W$ (see Sect. 4.3). Scale breaking due to QCD radiation should lead to a faster rise of the multiplicity, however, dampened by colour coherence and the running of α_s. A faster than logarithmic rise is observed when comparing the H1 data to lower W lepton–nucleon scattering experiments. The QCD model LEPTO somewhat overestimates the multiplicity (see Fig. 5.4). The DIS data are also compared to the JETSET prediction (light quarks only) for e^+e^- data, which is known to give a good representation of the e^+e^- data. The behaviour of the total multiplicity in the CMS current hemisphere in DIS is similar to e^+e^- annihilation data. Differences are found, however, for the distribution in rapidity; see below.

The growth of multiplicity with energy has been calculated via the QCD evolution of fragmentation functions for running α_s in NLO [101, 134], yielding

---

[1] Rapidity gap events had been excluded from this analysis.

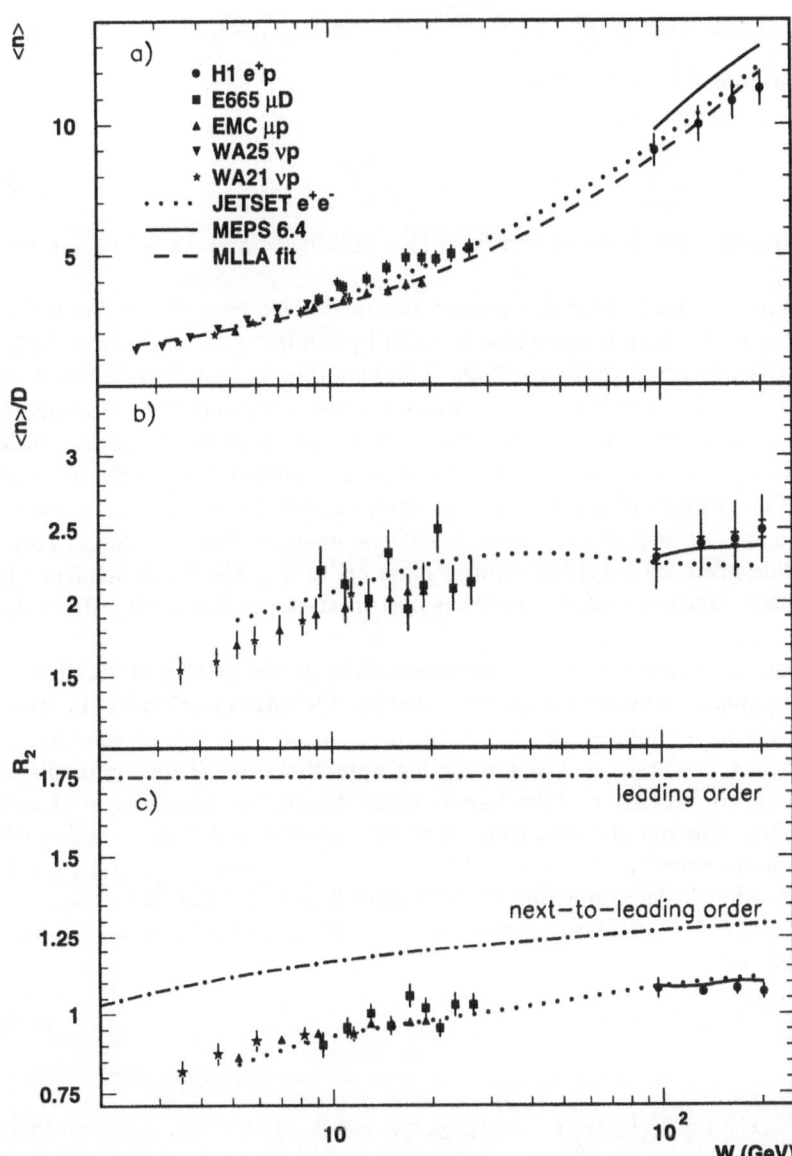

Fig. 5.4. (a) The average charged multiplicity $\langle n \rangle$, (b) the ratio $\langle n \rangle / D$, and (c) the factorial moment R_2 for charged particles in the CMS current hemisphere as a function of W. The DIS data are compared to the DIS QCD model MEPS, to the e^+e^- model JETSET for u, d, s quarks, and to a MLLA fit, and to the LO and NLO predictions for R_2

$$\langle n \rangle = a \cdot \left[\alpha_s(W^2) \right]^b \cdot e^{c/\sqrt{\alpha_s(W^2)}} \cdot \left[1 + d\sqrt{\alpha_s(W^2)} \right]. \tag{5.2}$$

The two-loop expression (2.8) for α_s is used with W^2 as scale. The constants b and c are given by the theory as

$$b = \frac{1}{4} + \frac{10 n_f}{27 \beta_0} \qquad \text{and} \qquad c = \frac{\sqrt{96\pi}}{\beta_0}, \tag{5.3}$$

with $n_f = 3$ active flavours and $\beta_0 = 11 - (2/3) n_f$. a and d are free parameters.

Predictions for hadron spectra can also be derived from the parton multiplicity in a jet, which is calculable in QCD by summing up all the branchings in the cascade down to a cut-off Q_0. The hypothesis of local parton–hadron duality (LPHD) (see Sect. 4.5) is made for the transition from partons to hadrons: the hadron spectra should have the same shape as the perturbative prediction, merely differing by a normalization constant. The prediction will depend on the cut-off Q_0, because the charged particle multiplicity is not an infrared safe observable, see Sect. 2.6. If one assumes that the shower stops at a scale given by a typical hadron mass $Q_0 \approx m_\pi$, the Q_0 dependence is eliminated. One can then compare the perturbative prediction directly to the data.

Most important for the average multiplicity is the growth at small fractional momentum due to soft gluon radiation. Quantum mechanical interference leads to a suppression of soft gluon radiation (soft colour coherence). The shower evolution in the leading-log approximation (LLA) is modified (hence called MLLA) to take into account destructive interference of soft gluons [92]. The modification consists of changing the evolution variable such that angular ordering [92, 135] between subsequent emissions i and $i + 1$ is fulfilled. That is, their opening angles satisfy $\theta_i > \theta_i + 1$ (see Sect. 4.4).

The MLLA+LPHD prediction for the hadron multiplicity for running α_s is [92, 100]

$$\langle n \rangle = c_1 \frac{4}{9} N_{LA} + c_2, \tag{5.4}$$

with

$$N_{LA}(Y) = \Gamma(B) \left(\frac{z}{2} \right)^{(1-B)} I_{1+B}(z), \tag{5.5}$$

where $z = \sqrt{\frac{48}{\beta_0} Y}$, $Y = \ln(W/2Q_0)$, $a = 11 + 2n_f/27$ and $B = a/\beta_0$. Γ is the Gamma function and I_ν the modified Bessel function of order ν. $Q_0 = 0.27$ GeV is used [100]. Equation (5.4) reduces to (5.2) for large z.

The data can be fit by either (5.4) or (5.2). For fixed α_s one would expect a faster growth than (5.4), namely a power law [100]

$$\langle n \rangle = a(W/W_0)^{2b'} - c, \tag{5.6}$$

which is still consistent with the data. In fact, all these fits would be indistinguishable in Fig. 5.4. The fit with (5.2) yields $a = 0.034 \pm 0.005$ and

Table 5.1. The parameters of fits to the dependence of the average charged multiplicity $\langle n \rangle$ in the CMS current hemisphere as a function of W [119]

ansatz	fit parameters		fixed parameters
(5.6)	$a = 1.40 \pm 0.04$	$b' = 0.20 \pm 0.01$	$c = 0.5, W_0 = 1$ GeV
(5.4)	$c_1 = 1.21 \pm 0.05$	$c_2 = 0.81 \pm 0.08$	$Q_0 = 0.270$ GeV
(5.2)	$a = 0.041 \pm 0.006$	$d = 0.2 \pm 0.3$	$\Lambda = 0.263$ GeV
(5.2)	$a = 0.034 \pm 0.005$	$\Lambda = 0.190 \pm 0.060$ GeV	$d = 0$

$\Lambda = 0.190 \pm 0.060$ GeV. In this fit the correction term $d\sqrt{\alpha_s}$ has been neglected, because it had been found insignificant ($d = 0.2 \pm 0.3$). It has to be pointed out that Λ extracted from (5.2) need not be identical with $\Lambda_{\overline{MS}}$, though they are expected to be similar if d is small [134]. The fit results are summarized in Table 5.1.

Multiplicity Distributions

H1 has also measured the multiplicity distribution P_n, giving the probability of observing n charged particles in a given event, as a function of W, Q^2 and for varying pseudorapidity ranges. In general, the data can be parametrized with the negative binomial distribution (NBD),

$$P_n(k, \overline{n}) = \frac{k(k+1) \cdots (k+n-1)}{n!} \left(\frac{\overline{n}}{\overline{n}+k} \right)^n \left(\frac{k}{\overline{n}+k} \right)^k. \qquad (5.7)$$

The two parameters $k, \overline{n}$ can be expressed in terms of the average multiplicity and the dispersion D:

$$D^2 = \langle (n - \langle n \rangle)^2 \rangle \qquad \langle n \rangle = \overline{n} \qquad \frac{D^2}{\langle n \rangle^2} = \frac{1}{\overline{n}} + \frac{1}{k}. \qquad (5.8)$$

One arrives at the NBD in many phenomenological models [136] for hadron production. A large body of data can be parametrized by the NBD. In QCD the LLA also leads to the NBD for the gluon multiplicity in a quark jet [137]. An alternative function is the lognormal distribution (LND), derived for multiplicative branching processes [138]. A variable is distributed according to the LND, if its logarithm is Gaussian distributed. The H1 data (Fig. 5.5) are equally well described by both parametrizations. The MEPS model deviates from the data both for very small and large multiplicities. In particular the origin of the excess of 0-prong events in the data over the MEPS model is not clear. For fixed W, a Q^2 dependence of $\langle n \rangle$ or D in the range 10 GeV^2 $< Q^2 < 1000$ GeV^2 was not found, within errors.

Hard gluon radiation would lead to a superposition of different NBDs, hence deviations from a single NBD for the hadron multiplicity distribution. A multi-jet induced shoulder structure as observed at LEP [139] is however not visible in the H1 data.

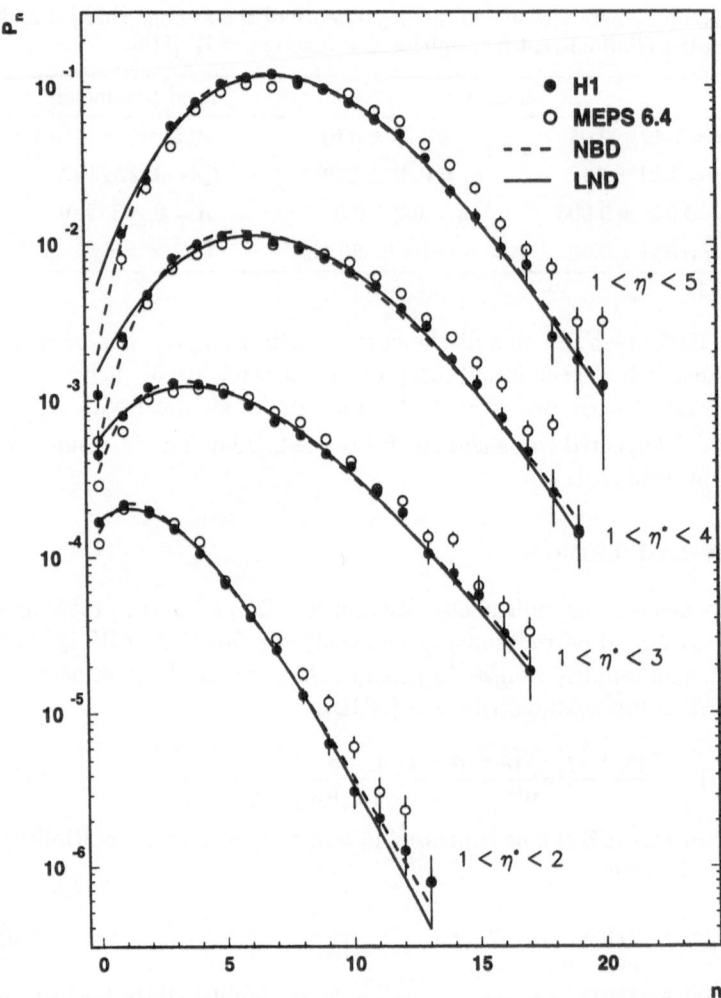

Fig. 5.5. The charged particle multiplicity distribution P_n for 115 GeV $< W <$ 150 GeV in indicated CMS pseudorapidity domains. The data [119] are compared to LEPTO (MEPS 6.4) and to the NBD and LND fit results. The data for $1 < \eta^* < 5$ are shown at the true scale, the others are scaled down by factors of 10 with respect to the previous one. Shown are statistical errors only

In Fig. 5.6 the multiplicity distribution is plotted in the scaling KNO (Koba, Nielsen and Olesen) form [140], $\langle n \rangle P_n$ as a function of $n/\langle n \rangle$. In a broad class of production mechanisms, based upon scale invariant stochastic branching processes, $\langle n \rangle P_n$ is expected to depend only on $n/\langle n \rangle$, and not on energy [141]. The H1 data for pseudorapidity $1 < \eta < 5$ for different W bins exhibit KNO scaling, and agree with the JETSET result for e^+e^- annihilation.

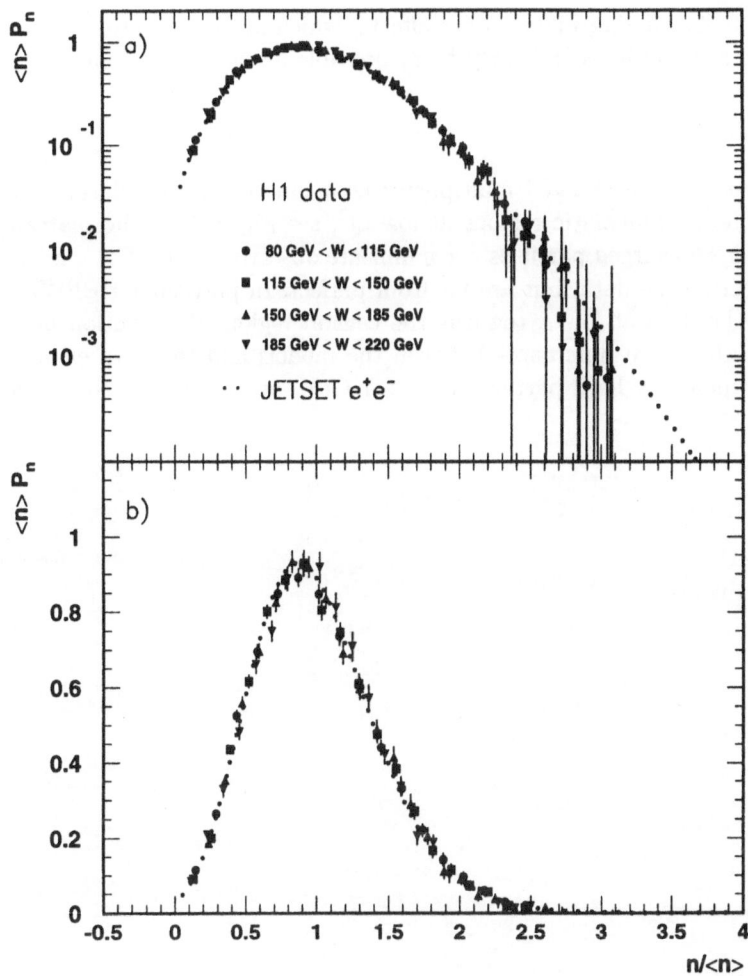

Fig. 5.6. The multiplicity distribution [119] for $1 < \eta < 5$ in KNO form for different W values, and compared to the JETSET prediction for e^+e^- annihilation. Shown are statistical errors only

In Fig. 5.4b and c the energy dependence for the ratio $\langle n \rangle / D$ and the normalized second-order factorial moment [142] $R_2 := \langle n(n-1) \rangle / \langle n \rangle^2$ are shown (further data on the moments of the multiplicity distribution can be found in [119]). For strict KNO scaling, they should be energy independent, which is the case for HERA energies at large W, but not for data at smaller W. The data are rather well reproduced by the JETSET e^+e^- model, and by the MEPS DIS model. A QCD calculation approaches the R_2 measurements when NLO corrections are taken into account [143], but still deviates significantly from the data. The LO result in the double logarithmic approximation (DLA), $R_2 = 7/4$, is reduced in NLO[143] to $R_2 = (1 - \chi\sqrt{\alpha_s}) \cdot 7/4$, where $\chi \approx 0.55$. Missing higher-order effects, which are taken into account with parton showers in MEPS, may be responsible for the disagreement.

Rapidity Distribution

H1 has measured the charged multiplicity as a function of pseudorapidity [121] for different kinematic regions at low Q^2; see Fig. 5.7. In the plateau region about 2.5 charged particles per η unit are observed. The QCD models roughly describe the data, but are far from perfect. In particular HERWIG overshoots the data at low x towards the central region. We mention here that quite substantial differences between the models and the data emerge at small x when only hard particles ($p_T > 1$ GeV) are considered (between

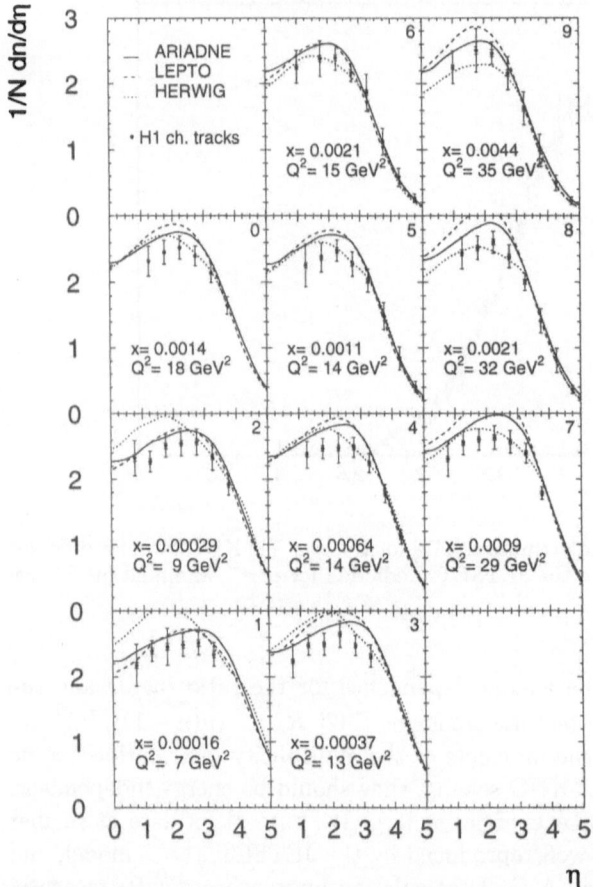

Fig. 5.7. The CMS pseudorapidity distribution of charged particles [121]. The proton direction is to the left. Data are shown for nine different kinematic bins with the mean values of x and Q^2 as indicated, plus the combined sample (bin 0). The models ARIADNE 4.08, LEPTO 6.4 and HERWIG 5.8 are overlayed

0.2 and 0.3 such particles are found per η unit). This effect will be discussed in detail in connection with low x physics in Sect. 8.3.

The charged particle density measured by H1 [119] close to the central region $1 < \eta < 2$, $\langle n \rangle \approx 2.5$ per η unit; is consistent with data from hadron–hadron interactions, see Fig. 5.8. When compared with other DIS experiments, a clear increase of the particle density with W is seen. The increase of the total multiplicity with W is not entirely due to the increasing longitudinal phase space (rapidity), but also due to an increase of the number of particles per unit rapidity, i.e. the height of the rapidity plateau increases with W. The average E_T at central rapidity (see Sect. 8.2) in DIS and in hadron–hadron collisions increases with a similar slope (Fig. 5.8)

The particle density at HERA is significantly lower than for e^+e^- data, represented by the JETSET result. Also the W dependence in e^+e^- is steeper than in hadron–hadron interactions. The DIS data appear to behave as one

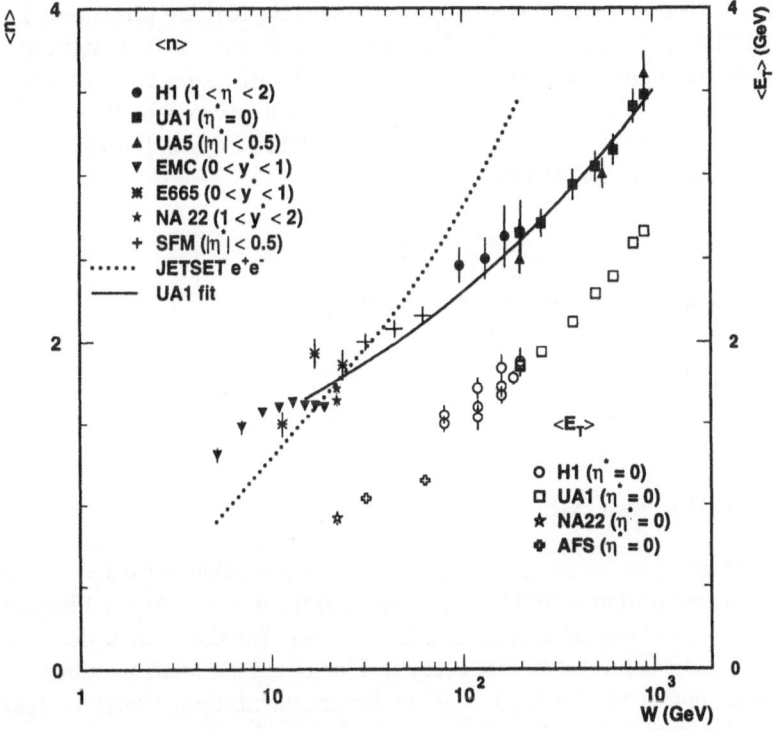

Fig. 5.8. *Solid symbols*: W dependence of charged particle density ($\langle n \rangle$ = # charged particles/unit pseudorapidity, left scale). *Open symbols*: W dependence of the mean central transverse energy $\langle E_T \rangle$ per unit pseudorapidity (right scale). The *dotted line* is the JETSET result for $\langle n \rangle$ in e^+e^- reactions, and the *full line* a parametrization $\langle n \rangle = 0.35 + 0.74(W^2)^{0.105}$ [145]. The DIS multiplicity data [119, 146, 147] and transverse energy data [133] are compared to hadron–hadron collisions [148, 149, 145, 150]. The DIS E_T data are discussed in Sect. 8.2

would expect from hadron–hadron data. That is also true for the transverse energy (see Sect. 5.1) measured at central rapidity [133]; see Fig. 5.8.

Why are there for the same CM energy W less charged particles in DIS than in e^+e^- at central rapidity? A possible explanation is the "antenna effect" [144], which leads to suppressed radiation from the smeared out colour charge of the proton remnant. Another possible explanation is that scaling violations leading to larger multiplicities are stronger in e^+e^- interactions than in ep reactions due to the larger scale, $Q^2_{e^+e^-} = W^2_{e^+e^-} = s_{e^+e^-} \gg Q^2_{ep}$. The x_F spectra in the CMS and their scaling violations are presented in the next section.

5.3 Charged Particle Momentum Spectra

In this section inclusive charged particle spectra measured by H1 and ZEUS [116, 151] with their central tracking chambers [116, 151] are presented. This limits the acceptance to the CMS current region, $x_F > 0$. Longitudinal and transverse momentum spectra are measured with respect to the virtual photon direction (Fig. 5.9). The kinematic region covered is $10 < Q^2 < 100 \text{GeV}^2$ and $55 < W < 200$ GeV for H1, and $10 < Q^2 < 160$ GeV^2 and $75 < W < 175$ GeV for ZEUS.

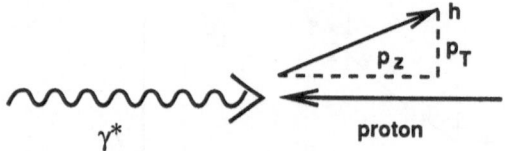

Fig. 5.9. The longitudinal (p_z) and transverse (p_T) momentum components of a hadron h in the $\gamma^* p$ CMS

x_F Spectra in the CMS

The Feynman-x ($x_F = 2p_z/W$) spectra are steeply falling with x_F, in fact more steeply than what would be expected from the naive QPM, that is quark fragmentation without QCD radiation (Fig. 5.10a). The data can be described when QCD radiation is taken into account. Due to gluon radiation, more than one parton shares the available energy for fragmentation; therefore fewer fast particles are produced, and the multiplicity of soft particles increases (Fig. 4.11).

The spectra can in principle depend on the available phase space, determined by W, and on the virtuality Q^2. A dependence on $x \approx Q^2/W^2$ is implied and important, because the parton composition of the proton changes with x. Table 5.2 gives the scales involved. The DIS x_F spectra measured at HERA [116, 151] are softer than the spectra measured at lower-energy fixed

ZEUS 1993

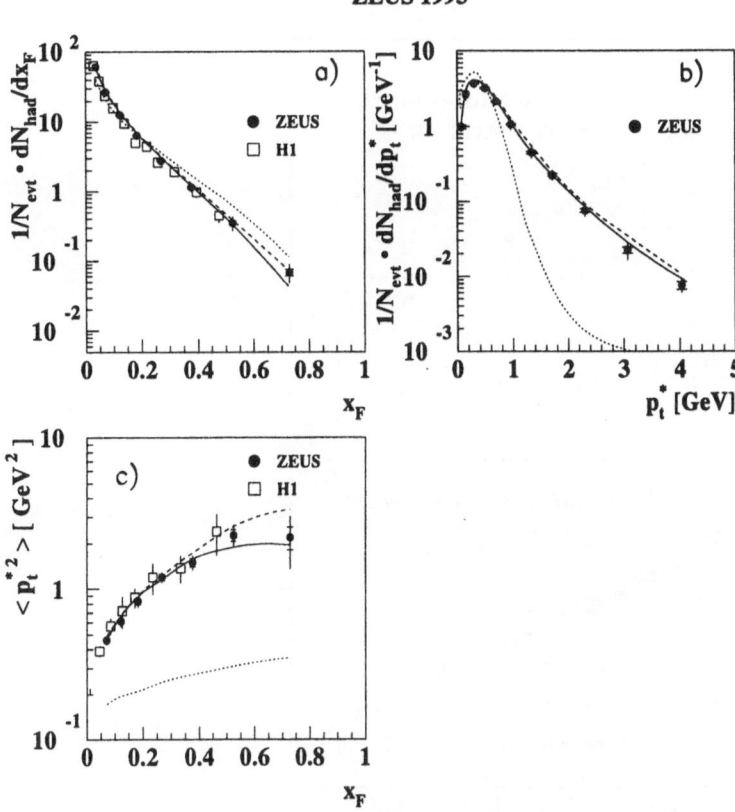

Fig. 5.10. Charged particle distributions in the CMS current hemisphere from H1 [130] and ZEUS [151]. (a) The Feynman x ($= x_F$) spectrum. (b) The p_T spectrum for $x_F > 0.05$. (c) The $\langle p_T^2 \rangle$ vs. x_F (seagull plot for $x_F > 0$). The data are compared with the QPM (*dotted line*) without QCD radiation, and the QCD models MEPS (LEPTO 6.1, *full line*) and CDM (ARIADNE 4.0, *dashed line*)

target experiments by EMC [152] and E665 [153]; see Fig. 5.11b. From this comparison it is, however, not clear whether the difference is due to the different values of x, W or Q^2.

Table 5.2. Approximate kinematics and scales for the e^+e^-, ep and fixed target lN data which are discussed here

	$\langle W \rangle$ (GeV)	$\langle Q^2 \rangle$ (GeV^2)	$\sqrt{s}$ (GeV)	$\langle x \rangle$
LEP e^+e^- ($\sqrt{s} = m_Z$)	91	8300	91	–
HERA ep	120	28	300	10^{-3}
lN (EMC/E665)	14/18	10	24/30	$5 \cdot 10^{-2}$

ZEUS 1993

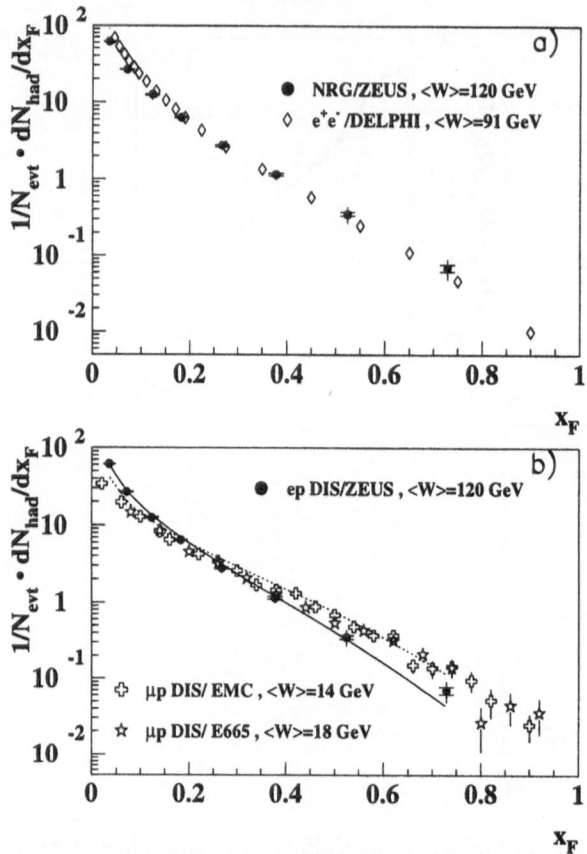

Fig. 5.11. (a) The x_F spectrum from ZEUS [151] compared with DELPHI data [154]. The DELPHI data are divided by 2. (b) The x_F spectra measured at HERA [116, 151], compared with the QPM (*dotted line*) without QCD radiation, the MEPS model (LEPTO 6.1, *full line*), and with fixed target lN data at lower W from EMC [152] and E665 [153]

In Fig. 5.11a the HERA data are compared with LEP data from DELPHI [154] (divided by 2 to account for the two hemispheres) at similar, though somewhat smaller centre-of-mass energy $\sqrt{s_{e^+e^-}} = W = m_Z$. For $x_F \gtrsim 0.1$ fragmentation univerality appears to hold, since the HERA and LEP spectra agree. At smaller x_F, the multiplicity is smaller in DIS than in e^+e^- interactions. (The different energy dependence of multiplicities at central rapidity in DIS and e^+e^- is discussed in Sect. 5.2, see Fig. 5.8.) The following circumstances may be held responsible for the difference: (1) mass effects due to the different flavour compositions; (2) the BGF contribution to DIS, which is absent in e^+e^- interactions; (3) the scale Q^2, which is smaller at HERA than at LEP; (4) the antenna effect due to the extended colour charge of the proton remnant (though $x_F = 0.1$ is already 2 units of rapidity away from the target region $\eta < 0$). Further studies are needed to identify the reason(s) for the difference.

A recent analysis by E665 [155] of the W and Q^2 dependence of the x_F spectra in the range $7.5 < W < 30$ GeV and $0.15 < Q^2 < 20$GeV^2 reveals that at low x_F they depend mainly on W, and at large x_F mainly on Q^2. The increase with W at small x_F is essentially due to the increase of longitudinal phase space. The decrease at large x_F has been attributed to the scale dependence of the fragmentation function.

QCD Predictions with Fragmentation Functions

The hadron production cross-section (see Fig. 5.12) $d\sigma/dz$ can be written as a convolution of the parton production cross-section with the fragmentation function $D_{h/j}(z, \mu_D^2)$, which gives the number density to observe a hadron h with momentum fraction z from the fragmentation of parton j. The parton production cross-section is itself a convolution of the parton density function $f_{i/p}(\xi, \mu_F^2)$ for parton species i in the proton with momentum fraction ξ, with the hard scattering cross-section $\hat{\sigma}_{ij}$ to produce parton j from electron scattering on that parton i, $e + i \rightarrow j + X$. Finally, the cross-sections need to be summed over all parton species i, j:

$$\frac{d\sigma}{dz} = \sum_{i,j} f_{i/p}(\xi, \mu_F^2) \otimes \hat{\sigma}_{ij} \otimes D_{h/j}(z, \mu_D^2). \tag{5.9}$$

The parton densities and the fragmentation functions depend upon their respective factorization scales, μ_F^2 and μ_D^2. The hard scattering cross section $\hat{\sigma}_{ij} = \hat{\sigma}_{ij}(\mu_F^2, \mu_D^2, \mu_R^2, \xi)$ depends in addition upon the renormalization scale μ_R^2 and α_s. The scales $\mu_R^2, \mu_D^2, \mu_F^2$ are in principle arbitrary; here they are taken to be Q^2 (not W^2!), the physical scale related with the hard scattering in DIS [156]. The situation in DIS with $Q \ll W$ is different to e^+e^-, where $Q = W = \sqrt{s} = 2E_q$.

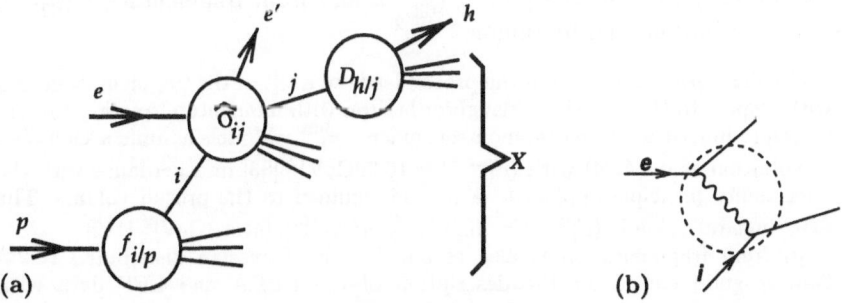

Fig. 5.12. (a) The process $ep \rightarrow e'hX$. In lowest order the blob labelled $\hat{s}_{ij}$ represents electron–quark scattering via the exchange of a virtual photon (b)

With experimentally determined parton densities and fragmentation function parametrizations from other, mainly e^+e^-, reactions [157], the NLO QCD calculation for the HERA x_F spectrum is quite satisfactory (Fig. 5.13) [156]. The LO calculation predicts a spectrum that is too hard compared with the data. When moving out of the current fragmentation region, i.e. towards small and negative x_F, these calculations are expected to break down due to fragmentation effects that are not related to the scattered parton.

Since the scaling violation of the parton density and fragmentation functions as well as the hard cross-section depend on α_s, a measurement of the scale dependence of $d\sigma/dz$ allows in principle a measurement of α_s. Due to the Q^2 range covered at HERA, it would be possible to measure α_s from the scaling violations of the fragmentation functions in a single experiment. For an integrated luminosity of 250 pb^{-1} a competitive statistical error for $\alpha_s(m_Z^2)$ of ± 0.0007 is expected. The theoretical uncertainties are presently estimated to be much larger [156]: ± 0.005 from the parton densities and ± 0.011 from the scale uncertainty. A better knowledge of parton densities would reduce the corresponding error. Possibly the theoretical situation is more favourable in the Breit system, where the scale of the hard interaction also sets the energy of the fragmenting quark, $Q/2 = E_q$, similar to e^+e^-. First indications of scaling violations in the Breit frame are presented in Sect. 6.1.

Transverse Momenta

The transverse momentum spectrum of charged particles from ZEUS with $x_F > 0.05$ is shown in Fig. 5.10b. The data exhibit a high p_T tail over the expectation from the QPM without QCD radiation, which is well described by the QCD models. p_T can be generated in the QCD cascade, where partons are emitted before and after the hard scattering. Somewhat arbitrarily one may distinguish radiation from the hard matrix element (BGF and QCDC) and soft gluon radiation (parton shower).

In general, contributions to the p_T of primary hadrons are expected also from other sources than QCD radiation (p_T^{rad}) [158, 5]: from an intrinsic p_T of the initial parton in the proton (p_T^{intr}), and from fragmentation (p_T^{frag}). These contributions can be summed up,[2]

---

[2] An intrinsic p_T of a parton in the proton leads to a tilt of the fragmentation axis with respect to the γ^* axis. A daughter hadron with momentum fraction x_F will therefore inherit a transverse momentum $x_F \cdot p_T^{\text{intr}}$. Models assume a Gaussian distribution as in (4.24) with $\langle p_T^{\text{intr}\,2}\rangle = (0.39\text{GeV})^2$ [83] in accordance with the uncertainty principle applied to a parton confined to the proton volume. The experimental value is $\langle p_T^{\text{intr}\,2}\rangle = [0.29^{+0.05}_{-0.07}(\text{stat.})^{+0.14}_{-0.18}(\text{syst.})\text{GeV}]^2$ [152].

p_T from fragmentation is also assumed to be Gaussian distributed (4.24) [94]. A good choice for the description of the HERA and LEP data with the Lund string model is $\langle p_T^{\text{frag}\,2}\rangle \approx (0.3\text{GeV})^2$ [83, 159, 126]. $\langle p_T^{\text{frag}\,2}\rangle = (0.41 \pm 0.02 \text{ GeV})^2$ has been measured with fixed target data [152].

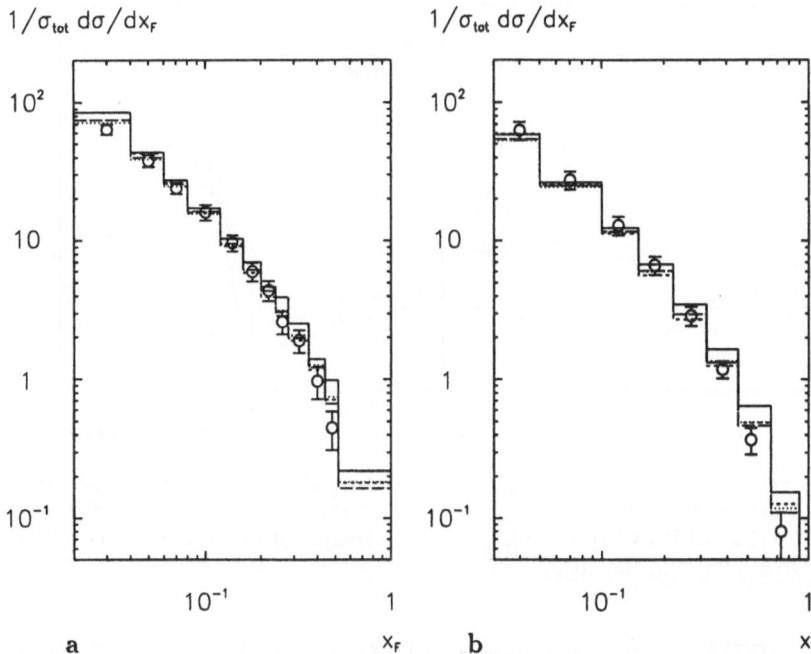

Fig. 5.13. The calculation by Graudenz [156] for different parton density parametrization compared to the HERA x_F data from (a) H1 and (b) ZEUS. The spectrum is calculated in LO with the GRV LO parametrization (*full line*), and in NLO with the parametrizations GRV HO (*dashed line*), MRSA′ (*dotted line*) and CTEQ3M (*long dashed line*)

$$\langle p_T^2 \rangle \approx x_F^2 \langle p_T^{\text{intr}\ 2} \rangle + \langle p_T^{\text{frag}\ 2} \rangle + \langle p_T^{\text{rad}\ 2} \rangle. \tag{5.10}$$

The p_T of the finally observed hadrons will be diluted due to decays of the primary hadrons. At HERA, the relative importance of p_T^{rad} is bigger than at the lower energy DIS experiments. This is best studied in the so-called seagull plot.

In the seagull plot the mean p_T^2 of charged hadrons is plotted as a function of x_F. The data from EMC [152], H1 [116] and ZEUS [151] are shown in Figs. 5.14 and 5.10c. The shape of this distribution is reminiscent of a seagull in the sky.[3] Due to the limited acceptance to $x_F \gtrsim 0$, at HERA only one wing of the seagull plot is measured.

---

[3] The dip at $x_F = 0$ is mainly a phase space effect. Consider massless hadrons distributed according to the longitudinal phase space model, $d^2\sigma/dydp_T^2 = a \exp(-p_T^2/b)$. Because $dy/dp_z = 1/E$, we have

$$\frac{d^2\sigma}{dx_F dp_T^2} = \frac{W/2}{E} \frac{d^2\sigma}{dydp_T^2}.$$

For small longitudinal momentum, $E = \sqrt{p_z^2 + p_T^2} \approx p_T$. So for $x_F \to 0$, the cross-section is weighted with a factor that grows with $1/p_T$ for small p_T, giving a large weight to small p_T hadrons in the average p_T^2.

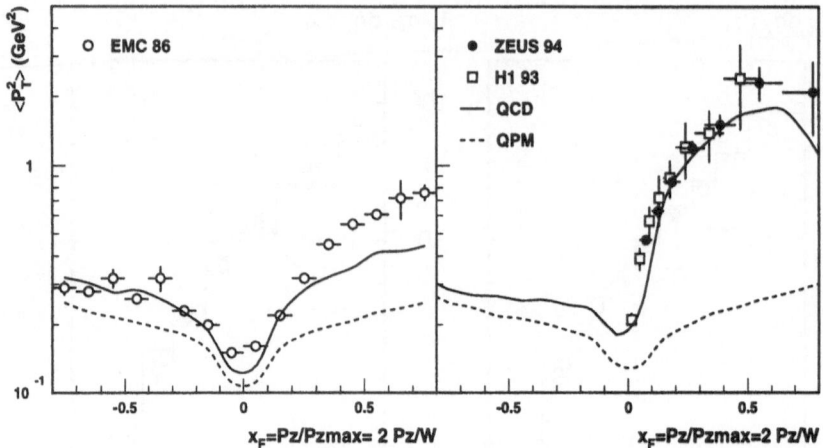

Fig. 5.14. The seagull plot, $\langle p_T^2 \rangle$ vs. x_F for charged hadrons in the CMS [160]. $x_F > 0$ defines the current hemisphere, $x_F < 0$ the target hemisphere. The data from H1 [116], ZEUS [151] and EMC [152] are compared to a QCD model (CDM, ARIADNE 4.08) and the QPM

The HERA data are well described by QCD models. Without QCD radiation (the QPM) the prediction falls far below the data. While the QPM predictions are quite similar for EMC and HERA energies, QCD radiation effects are much larger at the higher-energy experiments. The $\langle p_T^2 \rangle$ increases with W [151].

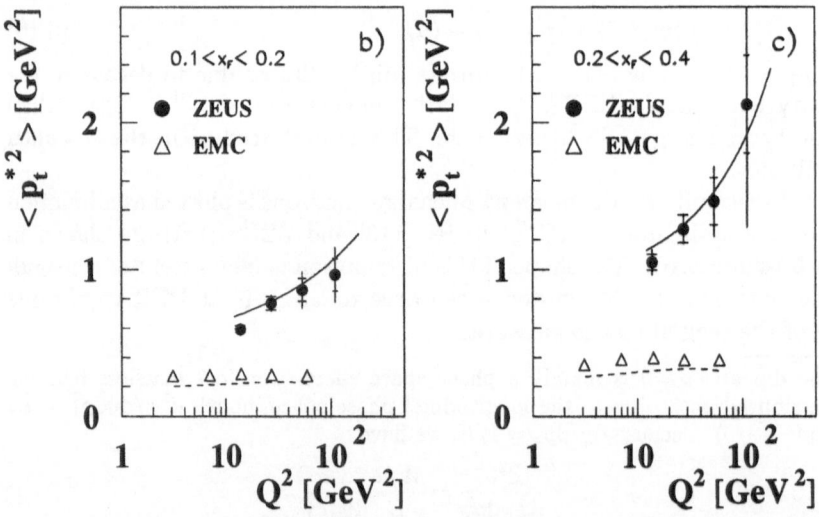

Fig. 5.15. The Q^2 dependence of the charged particle $\langle p_T^2 \rangle$ for different values of x_F. The ZEUS [151] data at $\langle W \rangle = 120$ GeV are compared to EMC data [152] at $\langle W \rangle = 14$ GeV. Also shown are the MEPS predictions for the two values of W

The fixed target data also cover the target fragmentation region, $x_F < 0$. In the target region QCD radiation is suppressed as compared to the current region. This suppression can be explained with the smeared out colour charge in the proton remnant, if one considers radiation from a colour dipole. Perturbative QCD evolution as implemented in LEPTO yields a similar asymmetry. The description of the EMC data by the CDM (ARIADNE 4.08) is certainly not satisfactory. It would be interesting to try to tune the QCD models to both data samples simultaneously.

At HERA energies an increase of $\langle p_T^2 \rangle$ with Q^2 is observed at fixed x_F, whereas at the lower energy fixed target data from EMC [152] almost no Q^2 dependence was found; see Fig. 5.15. With very precise data in the kinematic region $7.5 < W < 30$ GeV , $0.15 < Q^2 < 20$ GeV^2 and $1.5 \cdot 10^{-4} < x < 0.6$ the μp experiment E665 has recently seen a W as well as a Q^2 dependence for the mean p_T^2 [155].

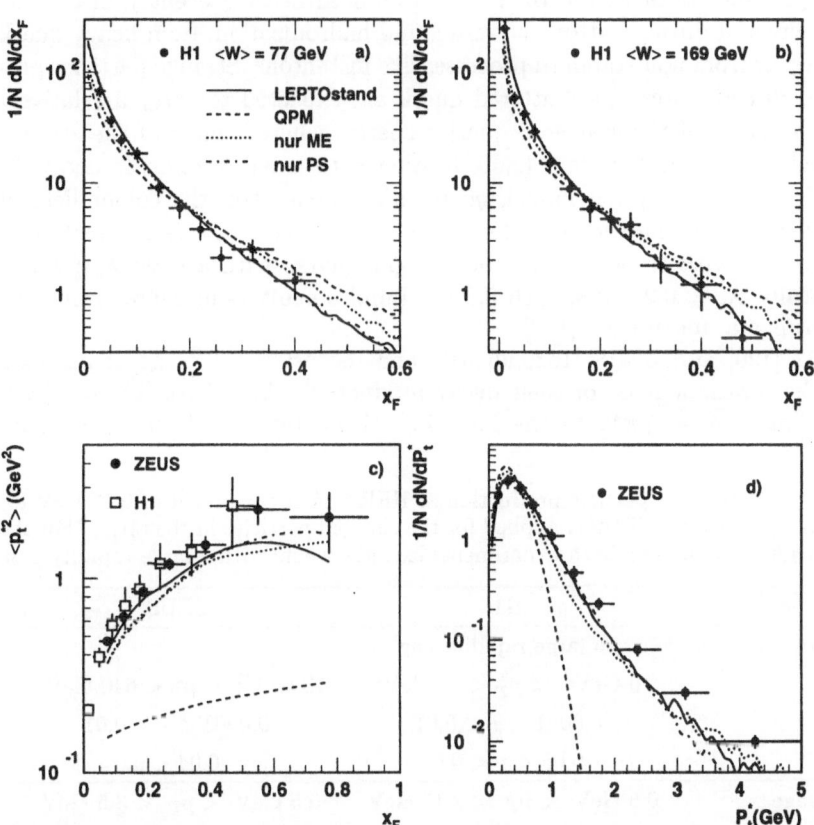

Fig. 5.16. QCD radiation effects according to the LEPTO 6.4 model [161]. The data are compared to the LEPTO model without QCD radiation (QPM), QCD radiation from the LO matrix element alone (*dotted line*, ME), QCD radiation from parton showers alone (PS, *dash-dotted line*), and the full model (*full line*)

We can use a QCD model to investigate the different contributions to the softening of the x_F spectrum and to the increasing p_T. In the MEPS model (LEPTO 6.4) radiation from the LO matrix element at the photon vertex (BGF and QCDC) can be identified as the main cause; additional soft gluon radiation is relatively unimportant (Fig. 5.16). On the other hand, the QCD parton shower as implemented in LEPTO appears to already cover a large part of the LO matrix element contribution. Note that the parton shower changes when the matrix element cut-off changes (or is switched off) in order to cover the maximal phase space without double counting (this adjustment is sometimes called "matching").

5.4 Strangeness Production

Strange particles may be produced in DIS when the scattered quark is a strange quark (either from the sea, or produced in BGF events), of from $s\bar{s}$ pairs created during parton showering and hadronization, from heavy quark decays, or from non-standard processes like instantons. Strange particles produced directly from the scattered quark are expected to carry a relatively large fraction of the scattered quark's energy, whereas most of the strange particles are expected from hadronization with small fractional energy. In the Lund hadronization model, $q\bar{q}$ pairs are created in the colour field of the string from the vacuum. Because $m_s > m_u \approx m_d$, $s\bar{s}$ pair creation is suppressed with respect to the other light quarks with a ratio $\lambda_s : 1 : 1$. Typically, $\lambda_s = 0.2 - 0.3$, with 0.3 the Lund default value representing the experimental mean from [162].

H1 [120] and ZEUS [163] identify and reconstruct K^0 and Λ particles via the invariant mass of their decay products in the decays $K_s^0 \to \pi^+\pi^-$, $\Lambda \to p\pi^-$, $\overline{\Lambda} \to \bar{p}\pi^+$. In the following, the notation K^0 and Λ includes

Table 5.3. Strange particle production at HERA. A lower p_T cut of 0.15 GeV for H1 and 0.2 GeV for ZEUS is applied for the charged particles in the ratio. *For the ratio #K^0/# charged both experiments exclude events with a large rapidity gap

	H1	ZEUS
event sel.	no large rapidity gap	
	$10 \text{ GeV}^2 < p_T^2 < 70 \text{ GeV}^2$	$10 \text{ GeV}^2 < p_T^2 < 640 \text{ GeV}^2$
	$0.0001 < x < 0.01$	$0.0003 < x < 0.01$
	$0.05 < y < 0.6$	$0.04 < y$
strange	$0.5 \text{ GeV} < p_T < 2.12 \text{ GeV}$	$0.5 \text{ GeV} < p_T < 3.5 \text{ GeV}$
particle sel.	$-1.3 < \eta < 1.3$	$-1.3 < \eta < 1.3$
# K^0/event	$0.287 \pm 0.008 \pm 0.012$	$0.289 \pm 0.015 \pm 0.014$
#K^0 / # ch. *	$0.058 \pm 0.002 \pm 0.003$	$0.077 \pm 0.006 \pm 0.008$
# Λ event	$0.053 \pm 0.007 \pm 0.007$	$0.038 \pm 0.006 \pm 0.002$

both particles and antiparticles. The decays are measured in the central drift chambers. This limits the acceptance for strange particles in the laboratory system to $-1.3 < \eta < 1.3$ and $p_T > 0.5$ GeV. The results for K^0 and Λ yields are given in Table 5.3. Neglecting the slightly different kinematic selections and averaging the ZEUS and H1 results, for $-1.3 < \eta < 1.3$ and $p_T > 0.5$ GeV the K^0 yield is 0.111 ± 0.005 per event and unit of pseudorapidity, and for Λ baryons it is 0.042 ± 0.0005.

The strange particle yield has been studied as a function of x, W, Q^2, η, p_T and x_F. In general, the same features are observed as for charged particles. These are an almost flat distribution in pseudorapidity, steeply falling p_T spectra, and no significant dependences on the event kinematics other than on W. The ratio of K^0 to charged particle yield was found to be the same in events with and without a large rapidity gap, within errors. The average K^0 multiplicity in the CMS current hemisphere increases logarithmically with W; see Fig. 5.17. This behaviour is similar to what has been seen for charged particles, though a faster than logarithmic rise is not seen with K^0 mesons (compare Sect. 5.2).

The data can be reasonably well described by standard QCD models invoking Lund string fragmentation, provided $\lambda_s \approx 0.2$ is chosen. As an

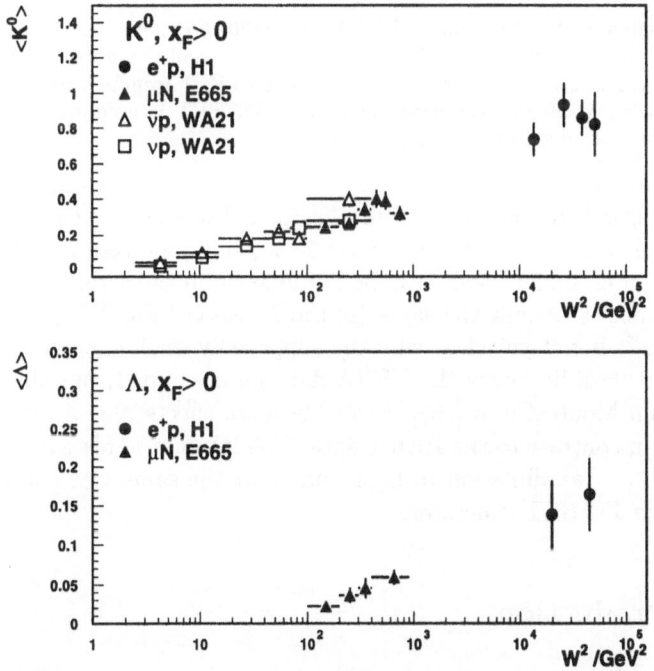

Fig. 5.17. The average K^0 (*top*) and Λ (*bottom*) multiplicity in the CMS current hemisphere as a function of W^2. The H1 data [120] are compared to fixed target data from E665 [164] and WA21 [165]

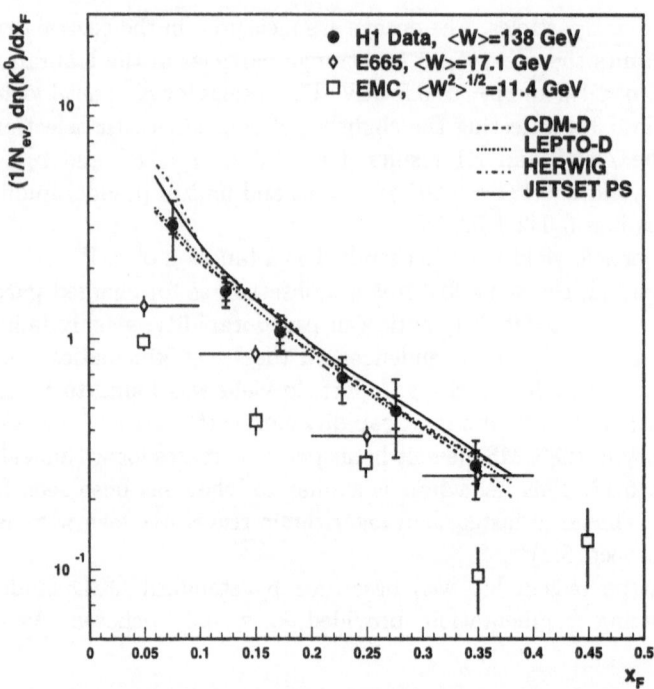

Fig. 5.18. x_F spectrum of K^0 mesons. The H1 data [120] are compared to fixed target data from EMC and E665 [166, 164], and to the DIS event generators CDM, LEPTO and HERWIG. For CDM and LEPTO the DELPHI fragmentation parameter set [167] with $\lambda_s = 0.23$ were used. The curve JETSET is a simulation of e^+e^- annihilation (u, d, s quarks only) at $\sqrt{s} = 138$ GeV

example, the x_F spectrum and the seagull plot for K^0 mesons are shown in Figs. 5.18 and 5.19. The larger $\langle p_T^2 \rangle$ observed when going from fixed target to HERA energies can be understood with QCD radiation in the same manner as for charged particles. About the same $\langle p_T^2 \rangle$ are observed for K^0s and for charged particles. It is not yet clear why the supposedly scaling x_F spectra of the fixed target data lie below the HERA data for $x_F > 0.1$, but this is also expected from Monte Carlo [168]. Probably mass effects play a rôle at smaller energies, in contrast to the HERA data. The H1 data agree with the expectation for e^+e^- annihilation to light quarks at the same CM energy, represented by the JETSET generator.

5.5 Charm Production

The Charm Yield

Charm production in DIS has been investigated by detecting $D^{*+} \rightarrow D^0 \pi^+ \rightarrow K^- \pi^+ \pi^+$ [169, 170] and $D^0 \rightarrow K^- \pi^+$ decays (and their charge conjugates).

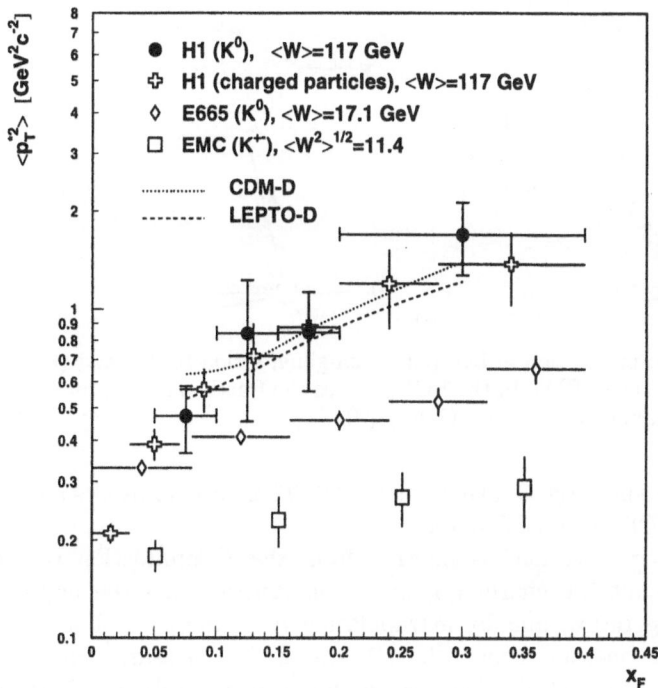

Fig. 5.19. The average p_T^2 as a function of x_F for K^0 [120] and charged particles [116] from H1, compared to fixed target data [166, 164] and QCD models

The decay products are detected in the central drift chambers. D^0 mesons are reconstructed by the $K\pi$ invariant masses. $D^\star$ mesons are identified by the kinematically tightly constrained mass difference $\Delta m = m(D^0\pi^+_{\text{slow}}) - m(D^0)$ between a $D^0 \to K\pi$ candidate and the slow pion, and the D^0 candidate mass.

Apart from possible anomalous sources of charm (new particles, instantons, ...), the main interest in charm production stems from its sensitivity to the gluon content of the proton at small fractional momentum x_g. The two lowest-order standard charm production processes in DIS are boson–gluon fusion (BGF) and scattering off a charm sea quark in the proton (QPM);[4] see Fig. 5.20. At high x also an intrinsic (valence) charm component to the proton at the few permil level has been discussed [171]. At small x the BGF contribution is expected to dominate due to the increasing gluon content. When the invariant mass $\sqrt{\hat{s}}$ of the $q\bar{q}$ system is large enough $(\hat{s} \gg (2m_c)^2)$ so that mass effects can be neglected, the quark flavours are expected to be produced according to $u : d : s : c = 4/9 : 1/9 : 1/9 : 4/9$ due to their electric charges, provided one is below the b threshold. $b\bar{b}$ production is suppressed by the large bottom mass. It has been estimated [172] that in DIS $b : c$ production is ≈ 0.02, integrated over a large range in x, Q^2, but can be as large

---

[4] The boundary between QPM and BGF is subject to definition.

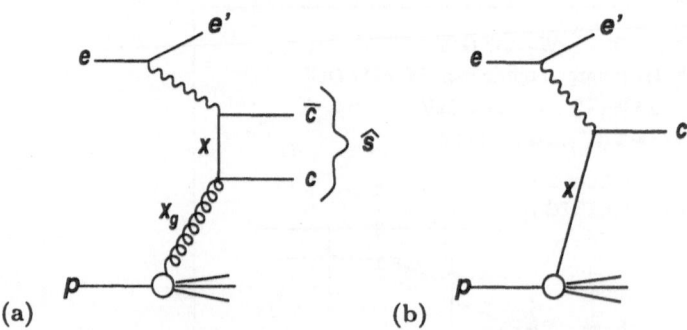

Fig. 5.20. Charm production in LO: (a) boson-gluon fusion (BGF) and (b) from sea or valence quarks (QPM). In the BGF process the fractional gluon momentum x_g is related to Bjorken x by $x_g = x(1 + \hat{s}/Q^2)$

as 0.1 at small x and large Q^2, that is large W. There are no results on open b production in DIS from HERA yet.

The charm cross-section[5] is inferred from the D production rate by taking into account the charm fragmentation function and the appropriate D branching ratios, and by extrapolating from the p_T and η regions accessible to the measurement. The H1 and ZEUS measurements cover $5 \text{ GeV}^2 < Q^2 < 100 \text{ GeV}^2$ and are summarized in Table 5.4. They are in reasonable agreement with each other, and with NLO calculations [174] based on current parton density parametrizations. About 4% of the DIS events contain a $D^{*\pm}$ [170], and about 25% of the DIS events are due to charm production [169, 170]. The asymptotic value of 40% expected from the quark charges is not yet reached. H1 finds for the ratio of $D^* : D^0$ production $0.38 \pm 0.07 \pm 0.06$, in agreement with other experiments (see compilation in [169]). Note that the D^0s include non-primary D^0s from D^* decays.

Table 5.4. Charm production at HERA [169, 170]. The extrapolation of the H1 measurement from the range $0.01 < y < 0.7$ to $0 < y < 0.7$ would modify the measured cross-section by approximately +6%. The NLO calculations [174], depending on the assumed gluon distribution function, and on the value of m_c used, vary between 9–13 nb and 8–14 nb for the two Q^2 bins

	H1	ZEUS
$\sigma_{c\bar{c}}$(nb)	$0.01 < y < 0.7$	$0 < y < 0.7$
$5 \text{ GeV}^2 < Q^2 < 10 \text{ GeV}^2$		$13.5 \pm 5.2 \pm 1.8^{+1.6}_{-1.2}$
$10 \text{ GeV}^2 < Q^2 < 100 \text{ GeV}^2$	$17.4 \pm 1.6 \pm 1.7 \pm 1.4$	$12.5 \pm 3.1 \pm 1.8^{+1.5}_{-1.1}$
	H1	ZEUS
	$8 \cdot 10^{-4} < x < 8 \cdot 10^{-3}$	$2 \cdot 10^{-4} < x < 5 \cdot 10^{-3}$
$\langle F_2^{c\bar{c}}/F_2 \rangle$	$0.237 \pm 0.021^{+0.043}_{-0.039}$	$\approx 25\%$

[5] The elastic (diffractive) cross-section for $ep \rightarrow epJ/\psi$ is comparably small, $100 \pm 20 \pm 20$ pb for $Q^2 > 8 \text{ GeV}^2$ and $30 \text{ GeV} < W < 150 \text{ GeV}$ [173].

Momentum Spectra of D Mesons

In order to investigate further the charm production mechanism, momentum spectra of the D mesons in the hadronic CMS have been measured [169, 170]. The p_T spectrum is well described by the Monte Carlo generator AROMA [175] based upon the BGF process (see Fig. 5.21a), and also by NLO calculations [170]. The spectrum of the scaled momentum variable $x_D := 2|\boldsymbol{p}|/W$ falls with increasing x_D (see Fig. 5.21b), in agreement with the BGF based charm generator, and also with the NLO calculations [170]. In contrast, QPM-like events from charm sea quark interactions would lead to an x_D spectrum that is peaked at $x_D \approx 0.6$, reflecting the hard charm fragmentation function. H1 excludes at 95% C.L. a 5% contribution from charm sea quarks to the total charm production.

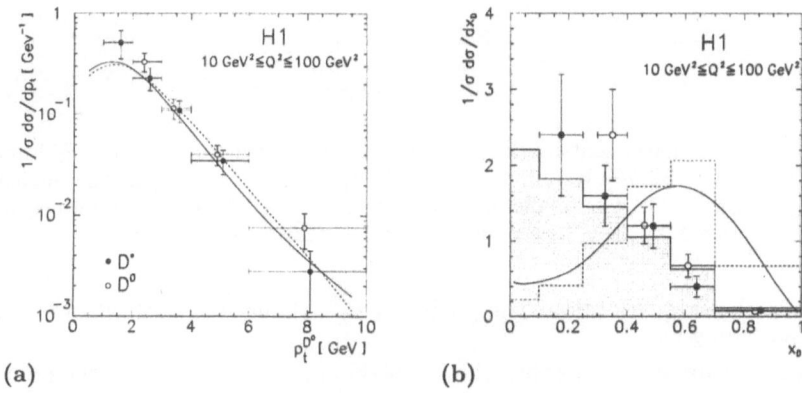

(a)   (b)

Fig. 5.21. (a) p_T spectrum of D^0 and D^* mesons [169], compared to the AROMA 2.1 [175] expectation for $m_c = 1.3$ GeV *(full line)* and $m_c = 1.7$ GeV *(dashed line)*. Only statistical errors are shown. (b) x_D distribution for D^0 *(open points)* and D^{*+} *(full points)* mesons in the laboratory pseudorapidity range $-1.5 < \eta_D < 1.5$. The data are compared to the expectation for BGF processes (AROMA, shaded), for QPM-like contributions from a charm sea (dashed histogram), and an extrapolation from charged current νN scattering off strange sea quarks [176] *(full line)*

The Charm Structure Function $F_2^{c\bar{c}}$

The cross-section measurements in bins of x and Q^2 can be expressed in the form of a charm structure function $F_2^{c\bar{c}}$ (Fig. 5.22) analogous to F_2. When compared with EMC data at $x = 0.02 - 0.3$, $F_2^{c\bar{c}}$ rises towards small x. The data are consistent with NLO calculations based exclusively on the BGF process with gluon densities that describe the current HERA F_2 data. In the calculation both the renormalization and factorization scales are chosen to be $\mu = \sqrt{Q^2 + 4m_c^2}$. The dominant uncertainty in the calculation arises

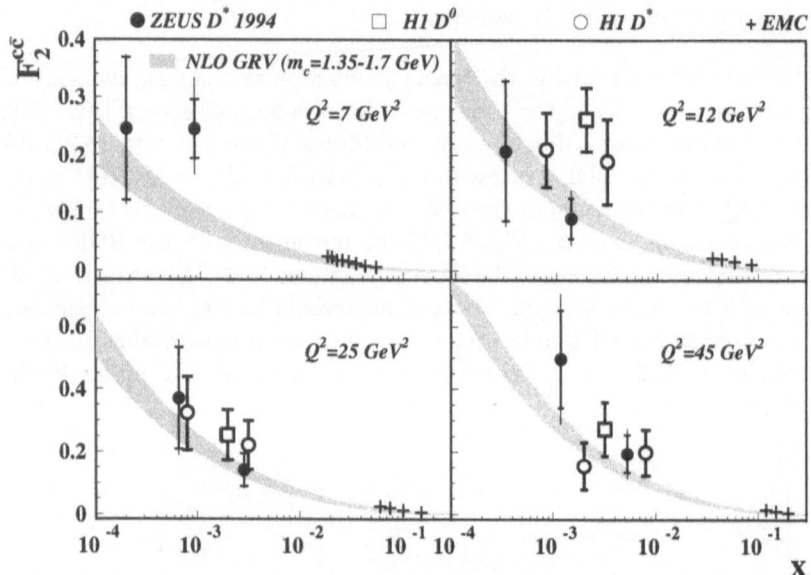

Fig. 5.22. The charm contribution $F_2^{c\bar{c}}$ to the proton structure function F_2 from H1 [169] and ZEUS [170]. Results from EMC [177] are shown as crosses. The shaded band gives the NLO QCD prediction from the GRV NLO gluon density for charm masses between 1.35 GeV (lower bound) and 1.7 GeV (upper bound)

from the charm quark mass. The predicted $F_2^{c\bar{c}}$ changes by 15% when m_c is changed by 0.2 GeV.

The measurements establish the boson–gluon fusion process as the dominant one for charm production in DIS at $Q^2 < 100$ GeV^2, and thus the sensitivity to the gluon distribution function in the proton. The data are consistent with parton density parametrizations which were derived from the inclusive F_2 measurements, providing an independent cross-check of the interpretation of the F_2 data in terms of parton densities.

The Gluon Density from Charm Production

In a new preliminary analysis [178] to determine the NLO gluon density from charm production H1 measures the visible $D^\star$ cross-section $\sigma_{\text{vis}}^{D^\star}$ (corrected for detector effects) in the range $2 < Q^2 < 100$ GeV^2, $0.01 < y < 0.7$ for $D^\star$ mesons with laboratory $p_T > 1.5$ GeV and $|\eta| < 1.5$. The measured cross-section is $\sigma_{\text{vis}}^{D^\star} = (5.63 \pm 0.66^{+0.84}_{-0.68})$ nb. The measured differential $D^\star$ cross-sections as functions of the kinematic variables x, Q^2, p_T, η (not shown) are well described by the AROMA [175] charm generator and by a NLO calculation (program HVQDIS 1.1 [179] with the Peterson charm fragmentation function [180]).

$x_{g/p}$ is the proton momentum fraction carried by the gluon entering the BGF process. $x_{g/p}$ can in leading order be reconstructed from

$$x_{g/p} = x(1 + \frac{\hat{s}}{Q^2}) \qquad (5.11)$$

with

$$\hat{s} = \frac{p_{Tc}^{*2} + m_c^2}{z(1-z)} \qquad (5.12)$$

and

$$z := \frac{P \cdot c}{P \cdot q} = \frac{(E - p_z)_c^{\text{lab}}}{2yE_e}. \qquad (5.13)$$

Here c is the charm quark's 4-vector and p_{Tc}^* its CMS transverse momentum. P is the proton 4-momentum, and E_e is the electron beam energy in the laboratory frame. The Lorentz invariant z can be calculated from the charm quark's energy E and longitudinal momentum p_z in the laboratory frame. For BGF events it can be correlated with the quark scattering angle in the photon–gluon CMS; see Sect 7.4 and (7.7).

Charm mesons are measured though not charm quarks. Therefore the observable x_g^{obs} is defined by replacing p_{Tc}^* with $1.2 \cdot p_{TD^*}^*$ and $(E - p_z)_c^{\text{lab}}$ with $(E - p_z)_{D^*}^{\text{lab}}$ in (5.13). x_g^{obs} is well correlated with $x_{g/p}$ [178].

The measured cross-section as a function of x_g^{obs} is shown in Fig. 5.23a. It is well described by the NLO calculation. The NLO calculation takes into account the gluon initiated processes $\gamma^* g \to c\bar{c}, c\bar{c}g$ and the quark initiated

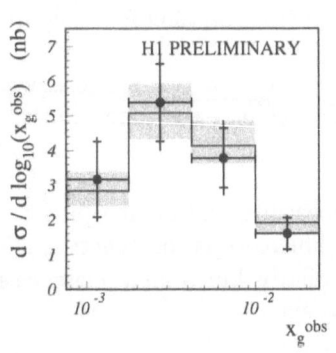

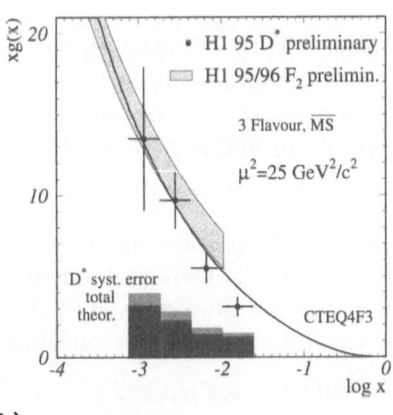

(a)   (b)

Fig. 5.23. (a) The visible D^* cross-section as a function of x_g^{obs} (see text), corrected for detector effects. The preliminary H1 data [178] (data points) are compared to a NLO QCD calculation [179] using the GRV 94 HO [49] parton density function (histogram; the shaded bands represent the variation of the cross-section when m_c is varied between 1.3 and 1.7 GeV). (b) The NLO gluon density $x \cdot g(x)$ at a scale $\mu^2 = 25$ GeV^2, unfolded from the D^* cross-section [178]. The statistical errors are shown as error bars, the systematic errors as shaded histograms. The line shows the CTEQ4F3 parametrization [57], the shaded band the gluon density from a NLO QCD analysis of the H1 F_2 data [41]

processes $\gamma^* q \to c\bar{c}q$, $\gamma^* \bar{q} \to c\bar{c}\bar{q}$, where the $c\bar{c}$ pair is produced from a radiated gluon. Using the NLO calculation for the x_g^{obs} cross-section, the gluon density can be unfolded from the data. In the calculation $m_c = 1.5$ GeV is used, and the factorization scale is set to $\mu_F = \sqrt{4m_c^2 + Q^2}$. The resulting NLO gluon density $xg(x, \mu^2)$ (Fig. 5.23b) at an average scale $\mu^2 = 24\text{GeV}^2$ reaches down to $x_{g/p} \approx 10^{-3}$, and agrees well with a recent indirect extraction from the F_2 data [41]. Smaller values of $x_{g/p}$ could be probed with a larger angular acceptance for D^* decays in the backward region, to be provided with silicon trackers. The ultimate limit is given by $x_{g/p} > \hat{s}/(sy) \approx 10^{-4}$.

By reconstructing event-by-event $x_{g/p}$, this measurement probes the gluon density locally in $x_{g/p}$. In contrast, with $F_2^{c\bar{c}}$ the charm contribution is measured as a function of Bjorken x, integrating over all $x_{g/p} > x$. The comparison of the directly measured gluon density with the one extracted from the inclusive F_2, and the agreement of $F_2^{c\bar{c}}$ with the prediction based upon such gluon distributions constitutes an important test of QCD with universal parton (here: gluon) densities. A similar test was obtained with dijet events in LO, see Fig. 7.5.

5.6 Bose–Einstein Correlations

Due to Bose–Einstein statistics, identical bosons prefer to occupy the same quantum state. In particle physics, the Bose–Einstein effect leads to an enhanced probability for identical bosons – like sign pions – to have similar momenta [181]. The shape and strength of their correlation function provide information on the production process.

As a measure of distance in momentum space for two particles with 4-momenta p_1 and p_2 with invariant mass $M = \sqrt{(p_1 + p_2)^2}$ one defines[6]

$$T^2 := -(p_1 - p_2)^2 = M^2 - 4m_\pi^2, \tag{5.14}$$

assuming the particles to be pions. One measures the number of like sign charged particle pairs as a function of T. This correlation function $\rho(T)$ is compared to a function $\rho_{\text{ref}}(T)$ obtained similarly, but from a reference sample where Bose–Einstein correlations are absent:

$$R(T) := \rho(T)/\rho_{\text{ref}}(T). \tag{5.15}$$

As reference sample (a) unlike sign charged particle pairs, or (b) pairs from different events (event mixed method) may serve.

If the particle-emitting sources are distributed in space–time according to $\rho_s(\xi)$, a ratio

$$R(T) = R_0(1 + \lambda|\tilde{\rho}_s(T)|^2), \tag{5.16}$$

---

[6] This variable is traditionally denoted as Q^2. Here we use T^2 instead, because in DIS Q^2 is already used for the virtuality of the exchanged photon.

is expected for the Bose–Einstein enhancement. $\tilde{\rho}_s(T)$ is the Fourier transform of $\rho_s(\xi)$ and R_0 a normalization constant. The correlation strength λ is 0 for completely coherent emitters, and 1 for completely incoherent emitters.

The measurement at small T, where the Bose–Einstein effect is expected, is notoriously difficult due to the finite double-track resolution of the tracking devices and due to e^+e^- pairs from photon conversions. The construction of a reference sample from the data introduces a residual bias. This effect is estimated by Monte Carlo, and cancelled by building the "double ratio"

$$RR(T) = R(T)_{\text{data}}/R(T)_{\text{MC}}. \tag{5.17}$$

The H1 data [182] are shown in Fig. 5.24. A clear enhancement at small T, attributable to the Bose–Einstein effect, is visible. The data cover $6 < Q^2 < 100 \text{ GeV}^2$, $1 \times 10^{-4} < x < 1 \times 10^{-2}$ and $65 < W < 240 \text{ GeV}$. No dependence of the Bose–Einstein effect on the kinematic variables x, Q^2 or W was found, nor a difference between the diffractive and non-diffractive data samples.

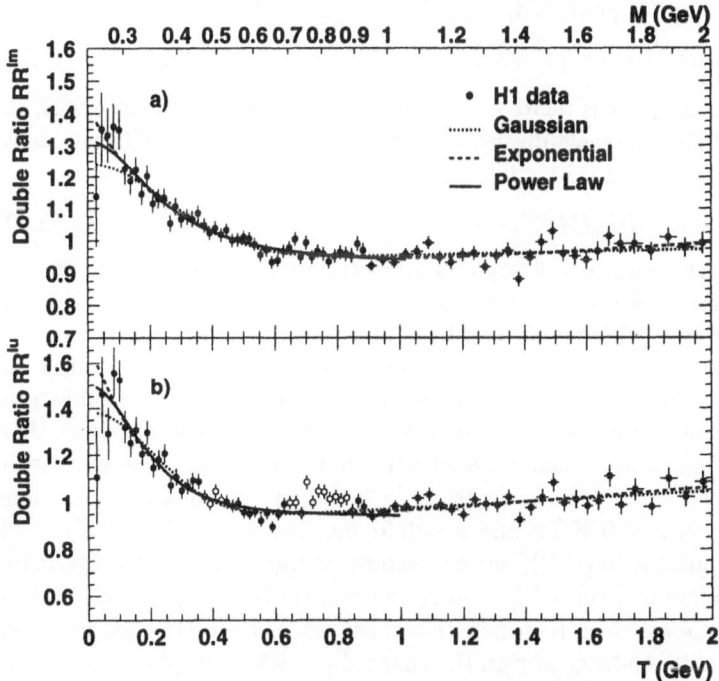

Fig. 5.24. Bose–Einstein correlations measured by H1 [182]. The double ratio $RR(T)$ is plotted as a function of T and M. For the reference sample, tracks from different events are paired (*top*), or unlike sign pairs are used (*bottom*). The fits of the Gaussian, exponential and power law parametrizations (with extra linear terms to accomodate the sloping background) are shown. The open points in the bottom plot are excluded from the fits, because they are in a region which is affected by correlations due to $K_s^0 \rightarrow \pi^+\pi^-$ and $\rho^0 \rightarrow \pi^+\pi^-$ decays

Table 5.5. The fit parameters from a Gaussian parametrization of the Bose–Einstein effect seen in the H1 data [182]

reference sample	r (fm)	λ
event mixed	$0.54 \pm 0.03^{+0.03}_{-0.02}$	$0.32 \pm 0.02^{+0.06}_{-0.06}$
unlike sign pairs	$0.68 \pm 0.04^{+0.02}_{-0.05}$	$0.52 \pm 0.03^{+0.19}_{-0.21}$

Different parametrizations of the effect have been fitted to the data. The traditional Goldhaber parametrization [183]

$$R(T) = R_0 \left[1 + \lambda \exp(-r^2 T^2)\right] \tag{5.18}$$

is derived from the the assumption of a static, spherically symmetric Gaussian source density $\rho_s(\xi) = \rho_s(0) \exp(-\xi^2/(2r^2))$. The parameter r gives the size of the production volume. In high energy reactions, the sources are moving relativistically [184], and the assumption of a static source becomes invalid [185]. In the relativistic string fragmentation picture [186], an approximately exponential shape is predicted,

$$R(T) = R_0 \left[1 + \lambda \exp(-rT)\right], \tag{5.19}$$

where the parameter r is related to the string tension [187]. Alternatively, assuming a self-similar, scale-invariant pattern for perturbative QCD cascades, a power law is derived [188, 142]

$$R(M) = A + B(1/M^2)^\beta. \tag{5.20}$$

Compared to the data (Fig. 5.24), all parametrizations are acceptable, though the exponential and the power law ansätze give somewhat better results at low T. A power law behaviour had been observed previously [189].

In order to compare with other experiments, the following results are obtained with the Gaussian (Goldhaber) parametrization (5.18) (Table 5.5). With the event mixed reference sample, a lower r value is obtained than with unlike sign pairs. Such a systematic effect has been observed also in other analyses. Other measurements in e^+e^- and lepton–nucleon interactions scatter between $r = 0.39$ fm and $r = 0.97$ fm, and $\lambda = 0.27$ and $\lambda = 1.08$ (see the compilation in [182]), but no pattern emerges. At least the extracted values of r, including the HERA ones, can be said to be in the ball park of the length scale at which hadronization takes place, 1 fm. H1 finds marginal evidence for a dependence of r on the charged particle multiplicity density $\bar{n}$, given as number of particles per unit pseudorapidity, $\bar{n} = dn/d\eta$. A strong multiplicity dependence of r, roughly $r = 0.4 + 0.075 \cdot \bar{n}$ (with $\approx 20\%$ accuracy) for $2 < \bar{n} < 20$ was seen in $p\bar{p}$ collisions [190, 191]. This is confirmed by the H1 data with a limited lever arm, $2 < \bar{n} < 5$.

6. The Quark Fragmentation Region

6.1 Charged Particle Spectra

Comparisons with e^+e^- Reactions

The physics issues that have been addressed with charged particle spectra in the Breit frame [192, 193, 194, 195, 196] are the study of fragmentation properties, QCD coherence effects, and scaling violations with the goal of determining α_s. The advantage of ep experiments over e^+e^- experiments at fixed beam energies is that the scale variable Q can be varied continuously. Complications arise however from initial-state parton radiation and from boson–gluon fusion events, which are absent in e^+e^-.

With the Breit current hemisphere a relatively small subsystem of the whole event is considered (for most events $Q \ll W$), which should be not much affected by the dynamics of the proton remnant (see Sects. 4.1 and 4.2 with Fig. 4.3). The Breit frame current hemisphere of an ep event, the quark fragmentation region, is similar to one hemisphere of an e^+e^- event. Considering just the lowest order process, in e^+e^- interactions the outgoing quarks have opposite momenta of equal magnitude. The same is true for the incoming quark and the outgoing quark in the Breit frame of DIS. Furthermore, as in e^+e^-, the scale of the hard interaction, Q, also determines the phase space for QCD effects, given by the quark energy $E_q = Q/2$. In contrast, in the CMS $E_q = W/2 \gg Q/2$. In the CMS it is therefore more difficult to disentangle a kinematic effect from a scale-dependent QCD effect.

The scaled momentum variables of a hadron with momentum p and energy E are defined as (compare Sect. 4.2)

$$ x_p := \frac{|p|}{p^{\max}} \qquad \xi := \ln \frac{1}{x_p} \tag{6.1} $$

with $0 < x_p < 1$. In practice, one often approximates $p^{\max} \approx E_q = Q/2$ (massless kinematics). With the variable ξ the soft part of the momentum spectrum at small x_p, where the particle density is the largest, is expanded.

Figure 6.1 shows the measured [192] x_p and ξ distributions of charged particles in the Breit current hemisphere for large and for small Q^2. $D(x_p)$ is steeply falling, and $F(\xi)$ is approximately Gaussian. A qualitative explanation for the Gaussian shape is given at the end of this Section. When

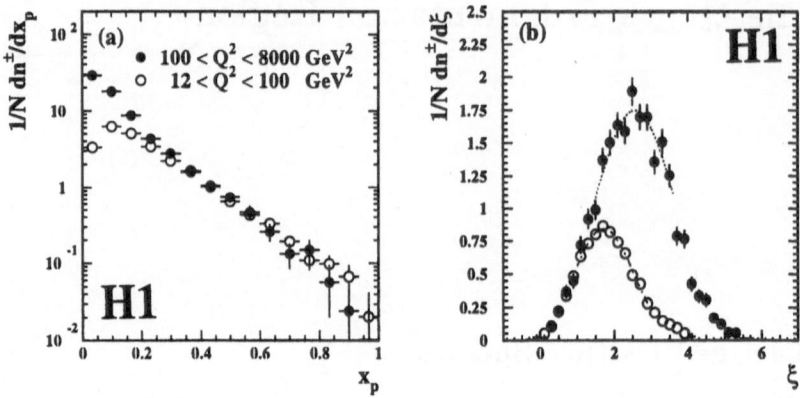

Fig. 6.1. The distributions $D(x_p) := 1/N \mathrm{d}n/\mathrm{d}x_p$ **(a)** and $F(\xi) := 1/N \mathrm{d}n/\mathrm{d}\xi$ **(b)** for charged particles from the Breit current hemisphere for large and for small Q^2 from H1 [193]. The curves are Gaussian fits to the data

Q increases, soft particles (large ξ) are produced much more copiously. The hard part of the spectrum (large x_p, small ξ) is approximately invariant of Q, "scaling". This can be understood naively. If hadron production stems from a branching process with energy-independent branchings, scaling at high x_p is expected. At low x_p, the reduction of the phase space forces a turnover of the spectrum, which happens at lower x_p for larger initial energy.

The evolution of $F(\xi)$ can be effectively summarized by the Q dependence of the peak position ξ_{peak} and width ξ_{width} of the Gaussian, shown in Fig. 6.2. The average charged multiplicity $\langle n \rangle$ (the integral over the fragmentation

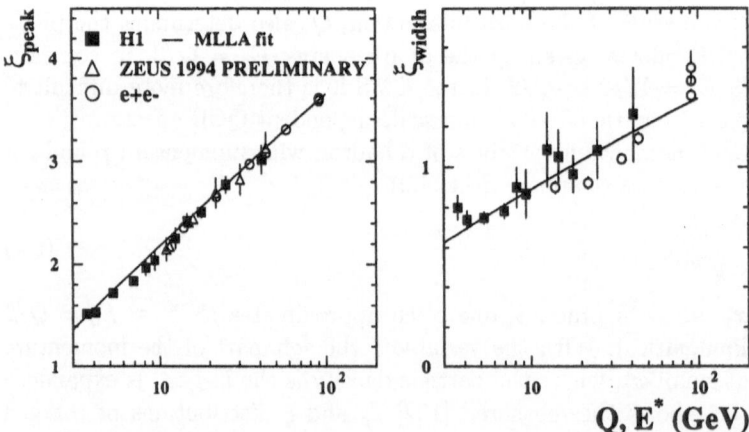

Fig. 6.2. The evolution of the peak position ξ_{peak} (*left*) and width ξ_{width} (*right*) of $F(\xi)$ with Q. Data from ep [193, 197] and e^+e^- reactions are compared. Also shown is a MLLA fit to the H1 data [193]

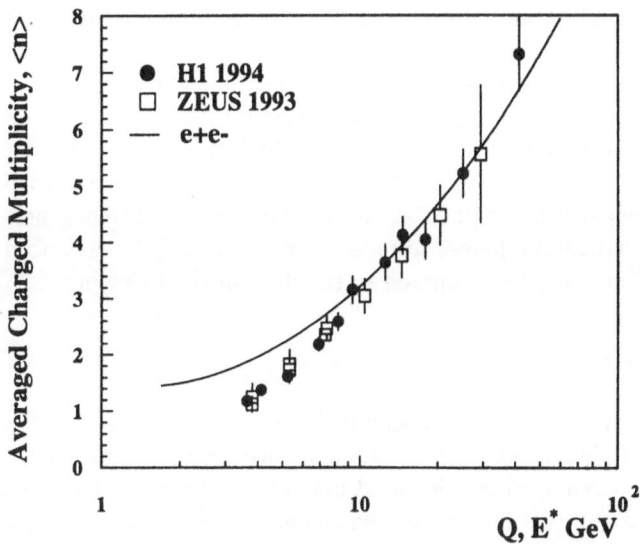

Fig. 6.3. Average charged particle multiplicity in the Breit current hemisphere as a function of Q. The H1 [193] and ZEUS [194] data are compared to a curve representing half of the charged multiplicity of e^+e^- annihilation events at CM energy E^*

function) is shown in Fig 6.3. An approximately logarithmic dependence of these variables on Q is observed:

$$\langle n \rangle \approx -1.7 + 4.9 \cdot \log_{10}(Q/\text{ GeV }) \tag{6.2}$$

$$\xi_{\text{peak}} \approx 0.6 + 1.5 \cdot \log_{10}(Q/\text{ GeV }) \tag{6.3}$$

$$\xi_{\text{width}} \approx 0.4 + 0.45 \cdot \log_{10}(Q/\text{ GeV }). \tag{6.4}$$

ξ_{peak} and ξ_{width} measured in e^+e^- annihilation at $\sqrt{s_{e^+e^-}} = E^* = Q$ are within errors consistent with the ep data. For $Q > 10$ GeV , also the total multiplicities $\langle n \rangle$ agree, but at smaller Q, $\langle n \rangle$ is significantly smaller in ep than in e^+e^-.

The deviation for $Q \lesssim 10$ GeV can be attributed largely to LO QCD processes in ep, in particular to the boson–gluon fusion process which occurs at small x (hence small Q) where the gluon density is large. For these events, the scattered quark may not be collinear with the photon, or may not even point into the Breit current hemisphere [198]. Its fragments are then emitted partly into the Breit target hemisphere.[1] For $Q \gtrsim 10$ GeV charged particle production in the ep Breit current hemisphere appears to be similar to one

---

[1] Consider the subsystem of the hard scattering subprocess with invariant mass $\hat{s}$ (QPM events with massless quarks have $\hat{s}=0$). One can easily verify that for $\hat{s} > Q^2$ the z momentum component of the hard subsystem is negative, pointing into the target hemisphere.

hemisphere in e^+e^- annihilation, supporting the notion of universal quark fragmentation. [2]

Predictions for the hadron spectra are available from (a) perturbative QCD in the modified leading log approxiation (MLLA), where the transition to hadrons rests on the hypothesis of local parton–hadron duality (LPHD); (b) perturbative QCD in NLO, based on the QCD factorization theorem with universal, but scale-dependent fragmentation functions; and (c) event generators with their sophisticated schemes to model perturbative QCD evolution and hadronization. They will be compared to the data in the following.

MLLA + LPHD Predictions

The MLLA+LHPD derivation of the spectrum $F(\xi)$ is discussed in [92, 199, 200]. The spectrum of partons in a parton cascade initiated by a quark of energy E_q (and related virtuality) is calculated perturbatively in the modified leading log approximation (MLLA), including colour coherence. The cascade is calculated down to a low cut-off $Q_0 \approx \Lambda_{QCD}$ for the parton virtuality. It is then assumed (the hypothesis of local parton–hadron duality, LPHD) that the final state hadrons follow the same spectrum, up to a constant factor. The predicted spectrum $F(\xi)$ has a "hump-backed" form of approximately Gaussian shape around the peak. On the other hand, when coherence effects are neglected, and $E_q \to \infty$ a plateau $F(\xi)$=const. should form in the soft region, until the kinematic limit is reached at a scale $\xi = \ln(1/x_p) \approx \ln(E_q/m)$. Coherence suppresses soft parton radiation, thus depleting the plateau at large ξ – a "hump-backed" plateau is formed.[3]

Analytical formulae in their various approximations for the spectrum are in general quite lengthy [92, 199, 200]. However, for large energies E_q and concentrating on the main part of the spectrum $F(\xi)$, without the tails, the peak and width of the Gaussian are predicted to evolve with E_q as [203]

$$\xi_{\text{peak}} = 0.5 \cdot \ln(E_q/\Lambda_{\text{eff}}) + c_2\sqrt{\ln(E_q/\Lambda_{\text{eff}})} + \kappa \qquad (6.5)$$

$$\xi_{\text{width}} = \sqrt{[\ln(E_q/\Lambda_{\text{eff}})]^{3/2}/(2c_1)}. \qquad (6.6)$$

The constants are $c_1 = \sqrt{36n_c/b}$, $c_2 = B\sqrt{b/(16n_c)}$ with $b = \frac{11}{3}n_c - \frac{2}{3}n_f$ and $B = \frac{1}{b}(\frac{11}{3}n_c + \frac{2}{3}n_f/n_c^2)$; they depend on the number of colours, n_c, and the number of flavours, n_f (here $n_f = 3$). κ accounts for higher-order corrections and is expected to be of order 1. Λ_{eff} is an effective scale for the evolution.

---

[2] In e^+e^- data the tails of the distribution deviate from a Gaussian. In the ep data, no significant deviation from a Gaussian has yet been found.

[3] A flat plateau $dn/d\xi$=const. implies also a flat rapidity plateau, dn/dy=const., since $dy/dp_z = 1/E$. A hump-backed distribution $dn/d\xi$ therefore results in a depletion for $y \to 0$. A dip in the rapidity distribution has been measured also in e^+e^- annihilation [201, 202]. The observation of a dip for 2-jet like events in e^+e^- annihilation provided evidence for soft and collinear gluon radiation [201].

It is not a priori clear that the above MLLA+LPHD calculations for showering quarks can be applied to ep data, where BGF events play a rôle at small Q. A good combined fit to the H1 data is obtained though, yielding $\Lambda_{\text{eff}} = 0.21 \pm 0.02$ GeV and $\kappa = -0.43 \pm 0.06$, in agreement with an analysis of combined e^+e^- data, for which $\Lambda_{\text{eff}} = 0.21 \pm 0.02$ GeV and $\kappa = -0.32 \pm 0.06$ was obtained [203].

It should be noted that the existence of a hump-backed plateau, Fig. 6.1, does not *prove* colour coherence effects. At finite energy, a hump backed, approximate Gaussian shape for the hadron spectrum can also be obtained in Monte Carlo simulations without coherence and with an independent fragmentation model [199]. One may argue that for an incoherent shower the turn over of the plateau happens at a scale $x_p \approx m/E_q$. This point would define ξ_{peak}, hence $d\xi_{\text{peak}}/d \ln E_q = 1$, in contradiction to the slower growth of the data and of the MLLA prediction. However, in an incoherent shower model with string fragmentation, also a slower growth close to the data can be obtained [194]. Furthermore, in [199] it is demonstrated how string fragmentation can turn a flat plateau for partons into a hump-backed one for hadrons. String fragmentation appears to compensate for negligence in the perturbative evolution. On the other hand, the growth of the average charged multiplicity with Q is less well described by string fragmentation with an incoherent parton shower [194].

Invariant Energy Spectra and the Soft Limit

Based upon MLLA+LPHD, the spectra from fragmenting quarks have been calculated also in the Lorentz invariant form

$$\frac{1}{N} E \frac{dn}{d^3 p} \tag{6.7}$$

to study the small momentum limit [204]. The calculation agrees well with e^+e^- data. One salient feature is that for $p \to 0$ the calculated spectra do not depend on the energy of the fragmenting parton. The invariant spectra from H1 [193] are shown in Fig. 6.4. The particle momentum p is measured in the drift chamber. Its energy is set to $E = \sqrt{Q_0^2 + |\boldsymbol{p}|^2}$, following the presciption in [204], where the "particle mass" is given by the mass cut-off of the parton shower, taken to be $Q_0 = 0.27$ GeV ($\approx \Lambda_{\text{QCD}}$).

Comparison of the MLLA+LPHD prediction for fragmenting quarks, which has been shown to agree with e^+e^- data [204], with the ep data is hampered by the presence of BGF events. At low Q, the prediction is far off the H1 data [193], see Fig. 6.4a. That effect can be reduced by a special anti-BGF event selection [193], which is based on the calorimetric energy seen in the Breit current hemisphere. That selection reduces the discrepancy at low Q. At large Q, where the discrepancy is smaller, there is little change after that cut, see Fig. 6.4b. This study shows that in the ep Breit frame spectra there are fewer fast particles than in e^+e^- events at the same energy.

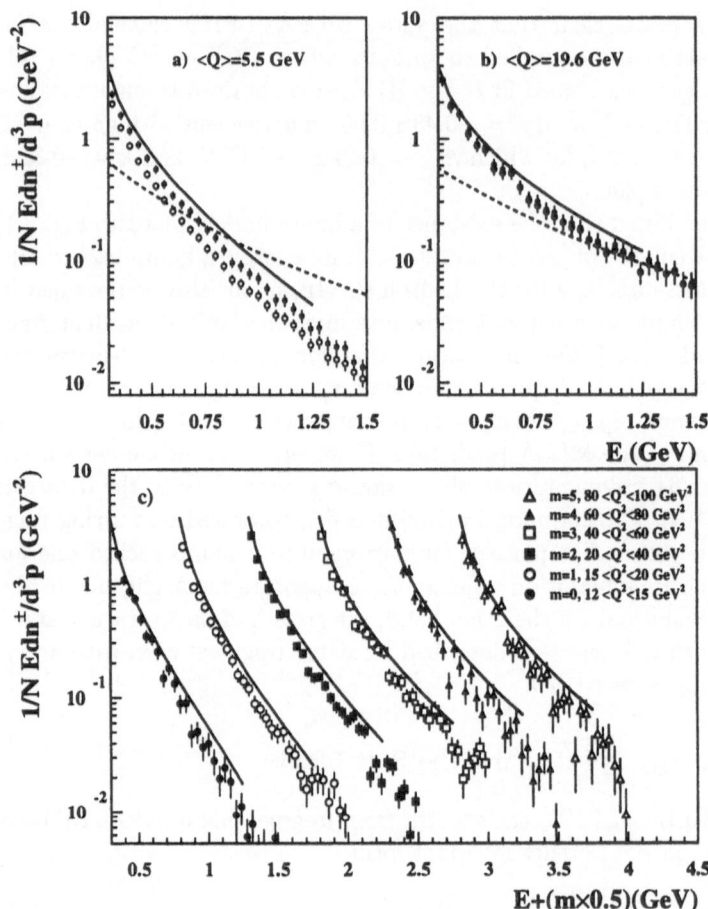

Fig. 6.4. Invariant charged particle spectra in the Breit current hemisphere [193]. In (a) and (b) the data at small and large Q are compared to the MLLA+LPHD prediction [204] with running α_s (*solid line*), and with constant α_s (*dashed line*). The open and full points are before and after the anti-BGF selection. In (c) the specta are shown for different Q^2 ranges. For better visibility, they are offset by an incremental spacing of 0.5 GeV . The *full lines* are the MLLA/LPHD expectations

The difference is presumably due to BGF events that are absent in e^+e^-. For larger Q, where the BGF contribution decreases, also the difference decreases.

The prediction of an energy (here: $E_q = Q/2$) independent soft spectrum is confirmed – the measured spectra converge for $|p| \to 0$, independent of Q (Fig. 6.4c). It can also be noted that a calculation with fixed α_s, instead of a running α_s, is clearly untenable (Figs. 6.4a, b). A meaningful extraction of α_s is however not yet possible, because the calculations are available only in LO.

In the same H1 analysis [193], the KNO multiplicity distribution was studied in the Breit current hemisphere. KNO scaling was found to be broken;

at low Q the distributions broaden, and deviate the most from the approximately scaling e^+e^- data. Again, BGF events and the limited rapidity range can be held responsible for this observation. It would be interesting to test also KNO-G scaling [205], which is a reformulation of the KNO form, valid for discrete distributions (like multiplicities) at finite energies.

Scaling Behaviour

The scaling behaviour of the x_p spectra in the Breit current hemisphere are best studied by displaying the Q^2 dependence of the normalized hadron cross-section $(1/\sigma_{\text{tot}})\mathrm{d}\sigma/\mathrm{d}x_p$ for fixed x_p, where σ_{tot} is the event cross-section. With the large range of Q^2 accessible at HERA, the scaling behaviour can be studied in a single experiment. Ultimately, α_s could be extracted from such measurements [156].

The ZEUS data [195], covering $10 < Q^2 < 1280\text{GeV}^2$, are shown in Figs. 6.5 and 6.6. The spectra were found to be almost independent of x for fixed Q^2 [192, 195]. Approximate scale independence is observed for $x_p \approx 0.3$: the cross-section $(1/\sigma_{\text{tot}}^{\gamma^*p})\mathrm{d}\sigma/\mathrm{d}x_p$ does not depend on Q^2. For smaller x_p, the cross-section increases strongly with Q^2 (compare also Fig. 6.1a). These effects are well described by the ARIADNE 4.08 [88] event generator; in LEPTO 6.5 [83] the scale dependencies are too strong, see Fig. 6.5. Possibly

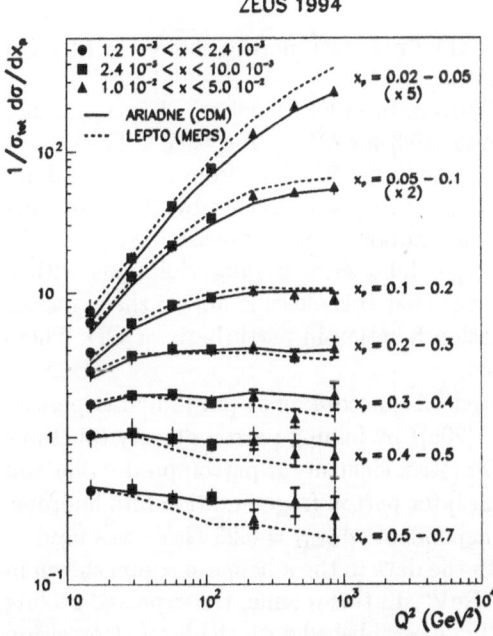

Fig. 6.5. Charged particle cross-section as a function of Q^2 for different x_p bins. The data [195] are compared to ARIADNE 4.08 [88] and LEPTO 6.5 [83]

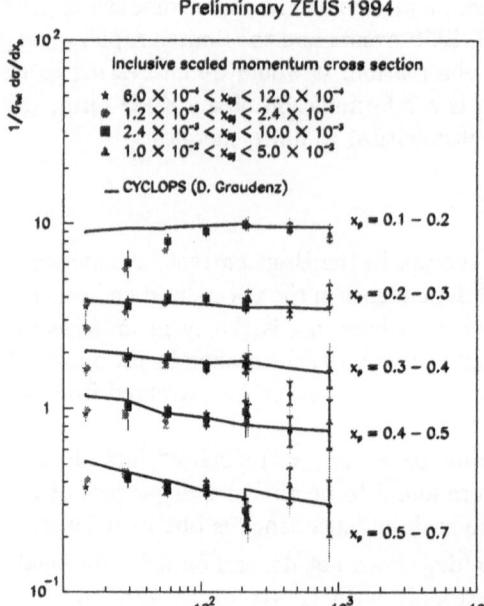

Fig. 6.6. Charged particle cross-section as a function of Q^2 for different x_p bins. The data [195] are compared to the NLO calculation with CYCLOPS [206]

the strong coupling parameter for the final state parton shower in LEPTO can be adjusted to describe the data.

In Fig. 6.7 new preliminary ZEUS data (1995) [196] are shown together with lower statistics H1 data (1994) [193] for $Q^2 > 100$ GeV^2. They agree very well. The ep data agree also very well with data from e^+e^- annihilation (corrected for K^0 and Λ decay products, and divided by 2), supporting the notion of universal quark fragmentation in this kinematic region. The measured cross-sections at small x_p exhibit clear scaling violations with a positive slope. It is not clear however that this effect is due to the expected scale dependence of the fragmentation function in perturbative QCD. These expectations will be discussed now.

The spectra have been calculated [211] according to (5.9) in NLO perturbative QCD (program CYCLOPS [206]) by folding parton density functions (here MRSA' [212]) with the NLO matrix element for parton production and an NLO fragmentation function [213] for parton fragmentation into hadrons. All of these ingredients are scale dependent. $\Lambda_{\rm QCD} = 0.23$ GeV was used.

The calculation agrees well with the data in the kinematic region shown in Fig. 6.7, $x_p > 0.1$ and $Q^2 > 100$ GeV^2. In this regime, the expected scaling violations are small, and can hardly be established with the present precision of the data. With higher precision data, and with a larger lever arm both at higher and smaller Q^2, the predicted scaling violations will be measurable. It will be very interesting to study to what extent this effect is due to the scale

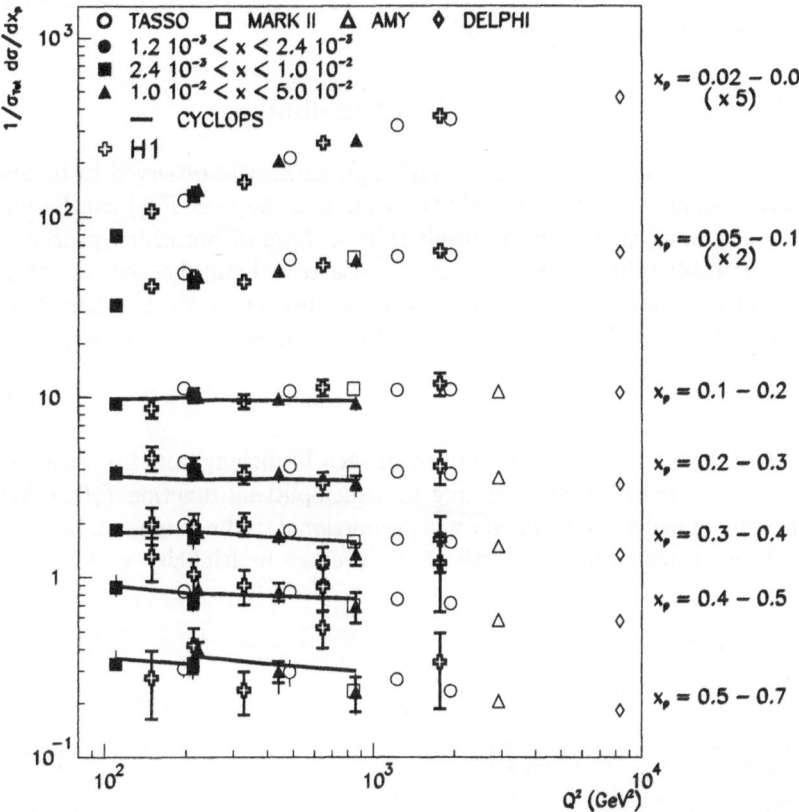

Fig. 6.7. Charged particle cross-section as a function of Q^2 for different x_p bins. H1 data [193] and preliminary ZEUS data [196] are compared to e^+e^- data [207, 208, 209, 210], and to an NLO calculation with the program CYCLOPS [156, 206]

dependence of the fragmentation function, the matrix element, or the parton densities.

The fragmentation function picture has its limitations [211]. The fragments must not be too far separated in rapidity from the parent parton, assumed to fragment independently. It is estimated that the hadron rapidity in the Breit system should be above some minimal value y', with $y' \approx 1$. The condition $y > y'$, translates into a lower bound on x_p,

$$x_p > \frac{2m_T}{Q} \frac{1}{\sqrt{1 - \tanh^2 y'}} \tag{6.8}$$

with the transverse mass m_T, typically of $\mathcal{O}(0.5~\text{GeV})$. Furthermore, hadron masses are not taken into account with fragmentation functions. The calculation thus must break down when $x_p \cdot Q/2$ approaches m_π. Indeed, the NLO calculation would by far overestimate the data in a region where

$Q^2 < 100$ GeV^2 and $x_p \approx 0.1$ [211], see Fig. 6.6. We note however deviations also at $Q^2 \approx 15$ GeV^2 and $x_p > 0.5$, where the NLO calculation should be reliable according to (6.8).

Gaussian Shape of the $\xi = \ln(1/x_p)$ Distribution

We return to the distribution of $\xi = \ln(1/x_p)$, which was observed to be approximately Gaussian. Qualitatively the Gaussian shape of $F(\xi)$ can be understood assuming that hadrons result from a chain of branching processes $i = 1, ..., k$ [214] (gluon coherence giving rise to a hump-backed, approximately Gaussian, shape is discussed later). The branching stops, when there is not enough energy left to produce a hadron of mass m. On average,

$$\frac{m}{E_q} \approx \langle z_i \rangle^{\langle k \rangle}. \tag{6.9}$$

The z_i give the energy fraction retained in each branching, and are assumed to be randomly distributed according to some splitting function $P_i(z_i)$. We expect that the hadron multiplicity n is proportional to the number of branchings k, hence a logarithmic growth of the average multiplicity with energy follows,

$$\langle n \rangle = c \cdot \ln(E_q/m). \tag{6.10}$$

The energy left after r branchings is

$$E_q^r = E_q \cdot z_1 \cdots z_r = E_q \prod_{i=1}^{r} z_i. \tag{6.11}$$

A hadron from this stage has energy $E^r = x_r \cdot E_q$ with $x_r = z_r' E_q^r / E_q$, where $z_r' = 1 - z_{r+1}$. We define the random variable $\xi_r := \ln(1/x_r)$, which is

$$\xi_r := \ln \frac{1}{x_r} = \ln \left(\frac{1}{z_r'} \cdot \prod_{i=1}^{r} \frac{1}{z_i} \right) = \ln \frac{1}{z_r'} + \sum_{i=1}^{r} \ln \frac{1}{z_i}. \tag{6.12}$$

ξ_r can be written as a sum of random variables, and is thus Gaussian distributed for $r \to \infty$ according to the central limit theorem, regardless of $P_i(z_i)$.

The final hadron spectrum of all hadrons from all k branching stages would then be the result of the sum of Gaussians $G_r(\xi)$ with $r > k$, where k itself is a random variable. For not too large k, this would give an approximate Gaussian, resulting from fall-offs at the kinematic boundaries at small and large ξ. For large k (large energy), a plateau $F(\xi) = $ const can develop, which translates into a rapidity plateau $dn/dy = $ const.

6.2 Event Shapes

Definition of Event Shape Variables

Information on the shape of events – pencil-like, spherical, cigar-like, planar, etc. – is conveniently given by simple functions of the hadron momenta p_i, $F = \mathcal{F}(p_i)$. Measurements of such event shape variables provide information about both perturbative and non-perturbative aspects of QCD. Due to the large kinematic range covered, HERA is particularly well suited to study their scale dependence, for example on energy, and by that means disentangle the two contributions. Furthermore, provided the non-perturbative part can be estimated, the strong coupling constant α_s can be extracted.

First measurements at HERA of the event shape variables thrust T, jet mass parameter ρ_C, and jet broadening parameter B_C are reported by H1 [215]. The measurements are restricted to the Breit frame current hemisphere in order to avoid complications from the not so well understood region towards the proton remnant. The energy of the scattered quark in the Breit frame (in the QPM), $Q/2$, provides the scale for the process. Q values between 7 and 100 GeV are covered.

The variables are defined by

$$T := \max \frac{\sum_i |p_i \cdot n_T|}{\sum_i |p_i|}$$

$$B_C := \frac{\sum_i |p_{Ti}|}{2 \sum_i |p_i|}$$

$$\rho_C := \frac{M^2}{Q^2} = \frac{(\sum_i p_i)^2}{Q^2}. \tag{6.13}$$

For the thrust calculation, the unit vector n_T, defining the direction of a thrust axis, is varied[4] in order to maximize T, the normalized longitudinal momentum sum of all particles i. $T = 1$ for a collinear event configuration, and $T = 1/2$ for a spherical configuration. Thrust measurements are usually expressed in $1 - T$, so that low values correspond to pencil-like events, as for the other event shape variables.

In the H1 measurement calorimeter clusters are used in the sums (6.13), which do not have a one-to-one correspondence to incident particles. However, these event shape variables are infrared safe quantities (see Sect. 2.6), they do not change when an object in the sum is split: $\mathcal{F}(p_1, \cdots, p_i, \cdots, p_n) = \mathcal{F}(p_1, \cdots, zp_i, (1 - z)p_i, \cdots, p_n)$.

Thrust Measurements

Here we shall concentrate on the thrust measurements, see Fig. 6.8; the conclusions from the other shape variable measurements are similar. The mea-

---

[4] H1 has also measured the thrust for fixed thrust axis, defined by the virtual photon.

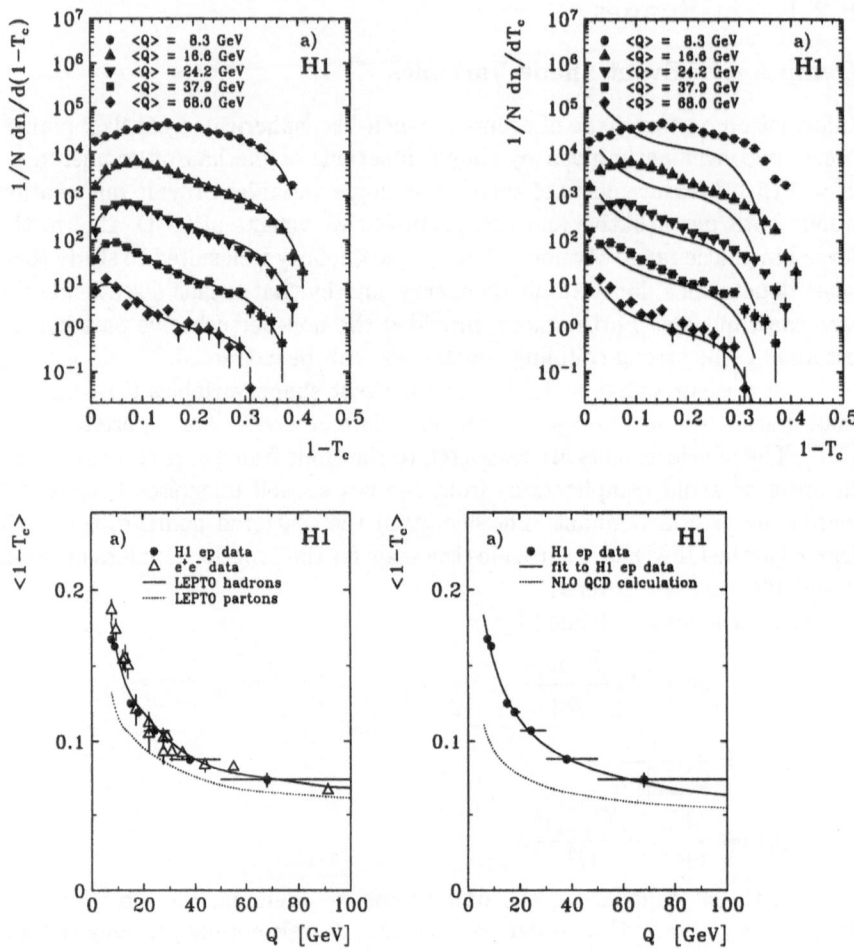

Fig. 6.8. *(Top)* thrust distributions [215]. The differential thrust distribution $(1/N)\mathrm{d}N/\mathrm{d}(1-T)$, compared to LEPTO 6.1 *(top left)* and an NLO calculation *(top right)*. The spectra for $\langle Q \rangle = 8.3\text{--}68$ GeV are multiplied by factors of $10^n, n = 0, ..., 4$. *(Bottom)* the mean $\langle 1 - T \rangle$ as a function of Q. The H1 data are compared to *(bottom left)* e^+e^- data for the full event (two hemispheres) at $\sqrt{s} = Q$, and the LEPTO result for hadrons and partons; and *(bottom right)* to the NLO calculation, and the fit result including the power correction

sured mean value $\langle 1 - T \rangle$ decreases with increasing Q; the events become more collimated. The thrust distribution as well as the mean thrust and their energy dependences are well described by the LEPTO model (version 6.1) [83], which incorporates a perturbative phase with the LO QCD matrix elements and leading log parton showers, and the Lund string model for hadronization.

Qualitatively, data from e^+e^- annihilation [216, 209, 207, 210, 208] follow the same trend as the ep data, but are systematically higher at small energies, $Q < 15$ GeV. One would actually not expect exactly the same distributions, partially because in e^+e^- the full event with two hemispheres is used.[5] Furthermore, the flavour compositions differ, and in ep there are $O(\alpha_s)$ processes contributing, namely initial state QCD radiation and boson–gluon fusion, which are absent in e^+e^-. In fact, the other shape variables (ρ_C, B_C) show bigger differences than thrust between e^+e^- and ep data.

Power Corrections

It has been argued [103] that for any infrared safe variable the mean value (and higher moments) can be written as

$$\langle F \rangle = \langle F \rangle^{\text{pert.}} + \langle F \rangle^{\text{pow.}}. \tag{6.14}$$

The perturbative part $\langle F \rangle^{\text{pert.}}$ is calculable in fixed order perturbation theory, depending only on α_s and the choice of renormalization scale.[6] The only energy dependence comes from the running α_s, since there is no other intrinsic scale in perturbative QCD than Λ_{QCD}. QCD radiation patterns would be identical at all energies, if α_s were constant. If the shape variable is invariant against momentum scaling with a constant c (like T), $\mathcal{F}(p_i) = \mathcal{F}(c \cdot p_i)$, the distribution of the event shape variable depends on Q only through the running α_s:

$$\frac{\mathrm{d}}{\mathrm{d}Q} \frac{\mathrm{d}N}{\mathrm{d}F} = \frac{\partial}{\partial Q} \frac{\mathrm{d}N}{\mathrm{d}F} + \left(\frac{\partial}{\partial \alpha_s} \frac{\mathrm{d}N}{\mathrm{d}F} \right) \cdot \frac{\partial \alpha_s}{\partial Q}, \tag{6.15}$$

because the first term on the right-hand side vanishes. In ep reactions a Q dependence may arise however via the Q-dependent parton densities.

Higher orders and hadronization effects are collected in the "power correction term" $\langle F \rangle^{\text{pow.}}$, which according to [103] scales approximately $\propto 1/Q$ (in general $\propto (1/Q)^p$ with the power p). A plausibility argument for such a term is given below. In contrast to the perturbative term, this term does depend explicitly on the energy, because hadronization introduces a scale, given by hadron masses. The higher-order perturbative effects are calculated with running α_s down to a scale μ_I, with $\Lambda_{\text{QCD}} \ll \mu_I$ to avoid the divergence of α_s for $Q \to \Lambda_{\text{QCD}}$, and $\mu_I \ll Q$ to maximize the perturbative evolution. A conventional choice is $\mu_I = 2$ GeV . Below the scale μ_I, the calculation uses a constant effective coupling $\bar{\alpha}_0$, a free parameter, which has to be determined by experiment, and which depends on the choice of μ_I.

The NLO expressions with Q as renormalization scale are [103]

$$\langle F \rangle^{\text{pert.}} = c_1 \alpha_s(Q^2) + c_2 \alpha_s^2(Q^2) \tag{6.16}$$

---

[5] Note that the thrust for a system of two identical back-to-back jets equals the thrust of a single jet.

[6] For the purpose of extracting a meaningful, process independent α_s, the calculation has to be done in fixed order perturbation theory to at least second order (NLO).

$$\langle F \rangle^{\text{pow.}} = a_F \frac{16}{3\pi} \frac{\mu_I}{Q}$$

$$\times \left[\bar{\alpha}_0(\mu_I^2) - \alpha_s(Q^2) - \frac{\beta_0}{2\pi} \left(\ln \frac{Q}{\mu_I} + \frac{K}{\beta_0} + 1 \right) \alpha_s^2(Q^2) \right], \quad (6.17)$$

where $\beta_0 = 11 - (2/3)n_f$, $K = 67/6 - \pi^2 - (5/9)n_f$ and $n_f = 5$ flavours.[7] The constant a_F is predicted by the theory and depends on the event shape variable F under consideration (see Table 6.1). In the H1 analysis, the constants c_1 and c_2 are calculated in the $\overline{\text{MS}}$ scheme with the program DISENT [78]; for example, for $\langle 1 - T \rangle$ they are $c_1 = 0.384 \pm 0.033$ and $c_2 = 0.57 \pm 0.21$.

With the ansatz (6.14) a good fit to the H1 $\langle 1 - T \rangle$ data (and $\langle \rho_C \rangle$, $\langle B_C \rangle$ as well) is obtained (see Fig. 6.8) with α_s and $\bar{\alpha}$ as the only free parameters. The perturbative contribution decreases with energy Q due to the decreasing α_s. The power correction is substantial, but decreases from $\approx 50\%$ at $Q = 8$ GeV to $\approx 20\%$ at $Q = 70$ GeV. The prediction from LEPTO for partons is higher and closer to the data than the NLO calculation (see Fig. 6.8), because the LEPTO parton shower includes QCD radiation in principle to all orders in the leading log approximation. In the ansatz (6.14), the higher orders are part of the power correction term. Whereas the differential thrust distribution dN/dT is well described by the LEPTO model, the NLO calculation (covering states with up to three final state partons) is insufficient, producing too narrow events at small Q, but approaches the data for larger Q. This is not surprising, as the large NLO correction to the LO in (6.16) (NLO/LO = $\alpha_s\, c_2/c_1$) suggests important higher order corrections, which should decrease with increasing energy due to the running α_s.

From the thrust fit one obtains for the two free parameters $\bar{\alpha}_0(\mu_I^2 = 4\text{GeV}^2) = 0.497 \pm 0.005^{+0.070}_{-0.036}$, and $\alpha_s(m_Z^2) = 0.123 \pm 0.002^{+0.007}_{-0.005}$ for the strong coupling extrapolated to the Z mass.[8] Similar values are obtained from the other event shapes, see Table 6.1.[9] Assuming the power corrections to be universal (same $\bar{\alpha}_0$ for all shape variables), a combined fit to the thrust and jet mass data yields $\alpha_s(m_Z^2) = 0.118 \pm 0.001^{+0.007}_{-0.006}$ [215]; one has to admit however that the small combined statistical error disregards the fact that the data samples that enter the different distributions are the same, not at all statistically independent. To be on the safe side, one should quote 0.003 as the statistical error instead. The obtained value agrees with the world average

---

[7] It is not clear what should be used for n_f in this situation. The lowest values of $Q/2$ are above the charm threshold, the larger ones are above the bottom threshold. On the other hand, for the parton shower with many soft emissions only light flavours are expected to be active. In fact, in the multiplicity analyses in Sects. 5.2 and 6.1 $n_f = 3$ was used.

[8] Note that (6.17) is no longer a pure power law $\propto 1/Q$, but is modified by corrections up to $O(\alpha_s^2)$. A pure power law fit of $\langle 1 - T \rangle^{\text{pow}} = 2\lambda/Q$ would give a worse χ^2 than the "QCD improved" expression (6.17) [217].

[9] According to [103], the power correction term for the jet broadening parameter B_C should be $\propto (1/Q) \ln Q$. From a comparison of the H1 data with the DISENT NLO calculation the extra $\ln Q$ term cannot be supported [215].

Table 6.1. Results of the event shape analyses [215]. Following [103], different values for a_F are used for the jet mass parameter in ep and e^+e^-

Observable	a_F	$\bar{\alpha}_0(\mu_I^2 = 4 \text{ GeV}^2)$	$\alpha_s(m_Z^2)$
H1 DIS			
$\langle 1 - T \rangle$	1	$0.497 \pm 0.005^{+0.070}_{-0.036}$	$0.123 \pm 0.002^{+0.007}_{-0.005}$
$\langle B_C \rangle$	2	$0.408 \pm 0.006^{+0.036}_{-0.022}$	$0.119 \pm 0.003^{+0.007}_{-0.004}$
$\langle \rho_C \rangle = \langle M^2/Q^2 \rangle$	1/2	$0.519 \pm 0.009^{+0.025}_{-0.020}$	$0.130 \pm 0.003^{+0.007}_{-0.005}$
e^+e^- **data**			
$\langle 1 - T \rangle$	1	$0.519 \pm 0.009^{+0.093}_{-0.039}$	$0.123 \pm 0.001^{+0.007}_{-0.004}$
$\langle M_H^2/Q^2 \rangle$	1	$0.431 \pm 0.020^{+0.071}_{-0.030}$	$0.115 \pm 0.002^{+0.005}_{-0.003}$

[14] with an error of similar size as in other determinations (see Table 6.1 and [218]).

It is quite remarkable that the ansatz (6.17) gives consistent values for α_s, and within 20% the same constant $\bar{\alpha}_0$ for all investigated shape variables. Furthermore, similar values are being obtained in shape analyses of e^+e^- data, see Table 6.1. These encouraging results support the concept of universal power corrections, which deserves further investigation of its origin in QCD, and of its limitations. Besides new insights into hadronization, a better understanding would also improve the α_s measurements. Recently also "power corrections" for the event shape distributions, not just the mean values, have become available [219]. It would be very interesting to compare these calculations to the data.

Naive Model for Power Corrections

The power behaviour $\propto 1/Q$ can be made plausible [75] with the simple tube or longitudinal phase space model [220] for hadronization. In this model a colour-connected pair of partons with total energy Q produces two back-to-back jets of hadrons with energy $E_j = Q/2$ each, where the hadrons are distributed uniformly in rapidity y and according to a probability density function $\rho(p_T)$ in transverse momentum. The resulting hadron distribution function is then $\Phi(y, p_T) = n_0 \rho(p_T)$, where n_0 gives the number of hadrons per unit rapidity. The energy and longitudinal momentum of one jet are given by

$$E_j = \int_0^{y_{max}} \int_0^\infty n_0 \rho(p_T) E \, \mathrm{d}p_T \mathrm{d}y = \lambda \sinh y_{max} \tag{6.18}$$

$$p_{jz} = \int_0^{y_{max}} \int_0^\infty n_0 \rho(p_T) p_z \, \mathrm{d}p_T \mathrm{d}y = \lambda(\cosh y_{max} - 1) \approx E_j - \lambda, \tag{6.19}$$

where λ is related to the average hadron p_T by

$$\lambda := \int_0^\infty n_0 \rho(p_T) p_T \mathrm{d}p_T = n_0 \langle p_T \rangle. \tag{6.20}$$

Used were $E = m_T \cosh y$, $p_z = m_T \sinh y$, $m_T \approx p_T$ for light hadrons, and the quick convergence of $\cosh y$ towards $\sinh y$. For a pencil like jet $p_{jz} = E_j$ would have been expected; hadronization yields a negative correction of relative size λ/E_j, a simple power of the energy.

In this model we can calculate hadronization corrections to the otherwise vanishing $1 - T_j$ and jet mass M_j of the jet. According to the definitions (6.13),

$$1 - T_j = 1 - \frac{p_{jz}}{E_j} \approx 1 - \frac{E_j - \lambda}{E_j} = \frac{2\lambda}{Q} \tag{6.21}$$

$$\frac{M_j^2}{Q^2} = \frac{E_j^2 - p_{jT}^2 - p_{jz}^2}{Q^2} = \frac{E_j^2 - p_{jz}^2}{Q^2} \approx \frac{2E_j\lambda}{Q^2} = \frac{\lambda}{Q}. \tag{6.22}$$

The corrections are proportional to $1/Q$, and twice as large for thrust as for the scaled jet mass squared, in accordance with the constants a_F for ep in Table 6.1. The different values for a_F in e^+e^- derived in [103] appear somewhat counterintuitive. One has to take into account however that the structure of DIS and e^+e^- events is more complicated than just two separating colour charges.

We can estimate the order of magnitude of the parameter $\lambda = n_0 \langle p_T \rangle$ from the data. From the seagull plot at low energies, where QCD effects are small, and for $x_F \to 0$ where intrinsic k_T effects vanish, we estimate $\langle p_T \rangle \approx 0.3$ GeV , see Fig. 5.14. From measured hadron rapidity distributions one gets for the hadron density per unit rapidity $n_0 \approx 2.0$ [5], compare also Fig. 5.8. That yields $\lambda \approx 0.6$ GeV. In fact, a good description of e^+e^- event shape variables is obtained with power corrections parameterized with $\lambda = 0.5$ GeV [75]. The fitted $\overline{\alpha}_0$ from the thrust analysis yields for the $1/Q$ term in (6.17) a coefficient of similar order, $2\lambda \approx 1.7$.

7. Jets

7.1 Introduction

Amongst the spray of particles emerging from a high-energy reaction the human eye can recognize collimated subsystems of hadrons, so-called jets (Fig. 1.2). Jets are fascinating objects; they allow one to view high-energy quarks and gluons which are not observable as free particles due to confinement. As such they can be considered as the parton images imprinted on the hadronic final state. At short distances, partons can be treated as "asymptotically free"; their interactions can be calculated perturbatively. When the partons emerge from the confinement volume, they fragment into observable hadrons. At high energies, the hadrons will be collimated around the original parton direction. These hadron jets still carry information on the underlying partonic interactions.

Jets in DIS at HERA result from the scattered quark, and from additional QCD radiation either in the initial or the final state. Clean jet structures can develop due to the large available phase space (W up to 300 GeV). The main experimental challenge in jet studies is twofold: (1) to measure jet quantities like energy–momentum, charge, etc., and (2) to relate them to the corresponding parton quantities. A good correlation between partonic and hadronic jet quantities is vital to allow perturbative QCD to be tested.

Jets at HERA have been studied with different intentions. The early measurements established clear jet structures in DIS [221, 222]. Fragmentation properties of quark and gluon jets have been studied by measuring jet shapes [223, 224]. Measurements of jet rates allow one to study the partonic QCD mechanisms and to measure the parameters by which they are determined, like the strong coupling strength α_s [225, 226], or the density of gluons in the proton [227]. It turns out that parts of the phase space for jet production, in particular at small x and Q^2, is not yet well understood [228, 229, 230, 231]. The dedicated search for BFKL "footprints" with forward jets is covered in the chapter on low x physics (Sect. 8.4).

7.2 Jet Algorithms

For quantitative measurements jets need to be defined: a jet is the output of
a jet algorithm. A jet algorithm prescribes how to combine objects (partons,
hadrons, energy deposits, ...) close in phase space to jets. It hinges on a
resolution parameter below which jets cannot be resolved. The jet algorithm
needs to be implemented not only experimentally, but also in the theoretical
calculation for meaningful predictions. There exist several jet algorithms for
different applications.

Clustering algorithms, like the JADE [232] or the k_T algorithm [233], combine *all* final state particles into few jets, based upon a measure of distance.
In the JADE algorithm, all particles i, j with invariant mass

$$m_{ij}^2 = 2E_i E_j (1 - \cos\theta_{ij}) < y_{\text{cut}} \cdot W^2 \tag{7.1}$$

are merged to form new objects. (Experimentally, at HERA a pseudopar-
ticle carrying energy and momentum of the unobserved proton remnant is
introduced.) Here y_{cut} is the resolution parameter, and W serves as reference
scale. The final set of jets is obtained, when no further merging is possible.

In the k_T algorithm two objects are merged, if

$$2 \cdot \min(E_i^2, E_j^2)(1 - \cos\theta_{ij}) < k_{\text{cut}}^2. \tag{7.2}$$

Particles can be merged also with the remnant. At HERA either $k_{\text{cut}} = Q$
is chosen [234], or k_{cut} is set to a fixed value, for example $k_{\text{cut}} = 3$ GeV
[228]. For small angles, the merging requirement becomes $k_T < k_{\text{cut}}$, where
k_T is the transverse momentum of the less energetic object with respect to
the other one (hence the name k_T algorithm).

For the cone algorithm [235] a distance measure in η (pseudorapidity) $- \phi$
(azimuth) is introduced: $\Delta R = \sqrt{(\Delta\eta)^2 + (\Delta\phi)^2}$. All particles inside a cone
with radius R (at HERA usually $R = 1$) are combined to form a jet, if the
resulting jet E_T with respect to the proton axis is above a certain cut-off
(usually a few GeV). The cone axis is chosen such that the E_T inside the
cone is maximized.

The result of a jet algorithm depends on the way masses are treated
when combining objects (recombination scheme). Measured jets consisting of
more than one particle always have a finite invariant mass, while most jets
in NLO calculations are massless (they consist of just one massless parton).
For this reason it is often found that the final result (e.g. α_s) depends on the
chosen recombination scheme. A lot of effort is being spent on minimizing
such effects. For the cone algorithm it also matters in which order objects
are combined.

7.3 Jet Shapes

ZEUS has measured the internal structure of jets in DIS and has compared
them to jets in other reactions [223]. One expects gluon jets to be broader

than quark jets due to the larger colour charge of the gluon. In the laboratory frame, jets with $E_T > 14$ GeV and $\eta \in [-1, 2]$ were selected with the cone algorithm ($R = 1$) for events with $Q^2 > 100\text{GeV}^2$. In most cases the current quark jet recoiling from the scattered electron will be selected.

One measures the average fraction $\psi(r)$ of jet transverse energy $E_T(r)$ inside a cone of radius r around the jet axis as a function r, defined by

$$\psi(r) := \frac{1}{N_{\text{jets}}} \sum_{\text{jets}} \frac{E_T(r)}{E_T(R)}. \tag{7.3}$$

N_{jets} is the number of jets in the event sample. $\psi(R) = 1$ by definition. The jet shapes are corrected for detector effects to the hadron level.

It is found that the jet shapes narrow as E_T increases, as expected kinematically. (A jet will look narrower with increasing boost.) They show no significant η dependence. (In a preliminary analysis of 2+1 jet events in the Breit frame, H1 found that these jets become broader towards the remnant [224].) The gross shape features are reproduced by the models investigated, PYTHIA 5.7 [236], LEPTO [83] and ARIADNE [88], but none of them provides an accurate description of the data over the entire phase space.

On average, DIS jets are seen to be narrower than jets from photoproduction. This can possibly be explained by the larger fraction of gluon jets in photoproduction, mostly from resolved processes, where a quark from the photon scatters with a gluon from the proton, $qg \rightarrow qg$. (However, multiple interactions between the proton and photon remnants could also lead to a jet broadening in photoproduction [237]). A similar observation is made when comparing DIS jets to jets from e^+e^- annihilation [238], and to jets from $p\bar{p}$ interactions [239] with similar E_T, see Fig. 7.1. DIS jets and e^+e^- jets are predominantly quark jets and have the same universal shape within errors. They are narrower than the jets measured in $p\bar{p}$ interactions which contain a significant fraction of gluon jets.

7.4 The Strong Coupling α_s

Analysis Method

Considering the LO graphs for dijet production (Fig. 7.2), the QCDC Compton (QCDC) and boson–gluon fusion (BGF) graphs, it is clear that the rate of events with 2+1 jets (the +1 refers to the remnant jet which escapes largely unobserved down the beam pipe) depends on the strong coupling constant α_s, and on the parton densities in the proton. Inspired by the great potential to measure the behaviour of α_s over the large span of Q^2 accessible at HERA with high statistical precision, jet studies have focussed on the determination of α_s right from the beginning of HERA data analysis. For this purpose one assumes parton densities that are constrained by other data.

The kinematics of the hard subprocess in 2+1 jet events is depicted in Fig. 7.2. One defines the invariant

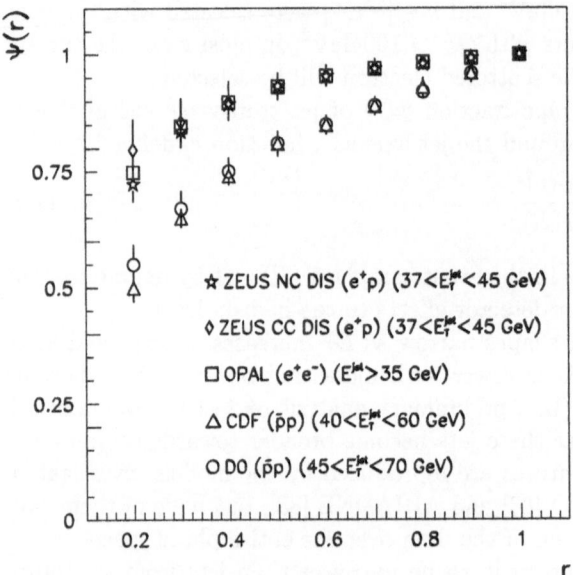

Fig. 7.1. The average jet shapes $\psi(r)$ for DIS charged (CC, $ep \to \nu X$ via W exchange) and neutral current (NC, $ep \to eX$ via γ, Z exchange) events [223], compared to e^+e^- jets [238] and $p\bar{p}$ jets [239]

$$\xi := x\left(1 + \frac{\hat{s}}{Q^2}\right) \approx \frac{\hat{s}}{W^2}, \tag{7.4}$$

where $\hat{s} := (j_1 + j_2)^2$ is the invariant mass squared of the 2-jet system. In LO ξ can be identified with the momentum fraction of the proton carried by the parton that enters the hard scattering, $x_{q/p}$ or $x_{g/p}$ for a quark or gluon.

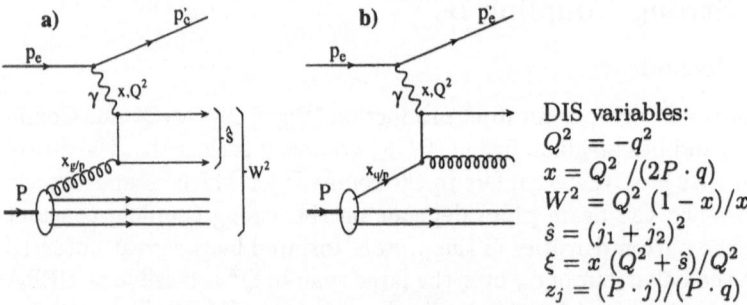

Fig. 7.2. a,b. Feynman diagrams for the production of $2+1$ jet events to first order α_s in ep-collisions. q and P denote the four-momenta of the photon and proton. j_1 and j_2 are the four-momenta of the jets associated to the hard subprocess with invariant mass squared $\hat{s}$. $x_{g/p}$ and $x_{q/p}$ denote the proton fractional momenta carried by the gluon or quark entering the hard subprocess

At present the strategy is to measure α_s with dijet data at relatively large Q^2, where the QCDC process dominates with well-known quark densities at large ξ. Dijet data at small Q^2, where the BGF process is the dominant one, are used to determine the less constrained gluon density at small ξ, assuming a value for α_s; see Sect. 7.5. Best use of the information in the data will in the future be made from a combined determination of α_s and the gluon density.

In the measurements the ratio of the number of 2+1 jet events N_{2+1} to the number of all events N_{tot},

$$R_{2+1} := \frac{N_{2+1}}{N_{\text{tot}}} \tag{7.5}$$

is determined as a function of Q^2. (H1 defines $N_{\text{tot}} := N_{1+1} + N_{2+1}$.) In the α_s analyses the JADE algorithm is applied in the laboratory frame with resolution parameter $y_{\text{cut}} = 0.02$. Another approach uses the dijet rate differential in the JADE resolution parameter y_{cut}, but no α_s value has been quoted yet [240]. R_{2+1} is corrected for (a) detector effects and (b) hadronization effects to the parton level,[1] and is then compared to the QCD prediction in NLO:

$$R_{2+1} = A \cdot \alpha_s(Q^2) + B \cdot (\alpha_s(Q^2))^2 + \mathcal{O}(\alpha_s^3). \tag{7.6}$$

A and B are calculable coefficients that depend on the kinematic conditions, the QCD renormalization and factorization scales, and on the parton densities. α_s is determined from a comparison of the measured jet rate, corrected to the parton level, with the NLO prediction of (7.6).

The analyses cuts are designed such that (a) the observed jet rates are well described by the Monte Carlo used for corrections; (b) the correction factors are reasonably small; (c) the NLO prediction for partonic jets resembles the partonic jets found in the Monte Carlo generator. Jets from initial state parton showers complicate matters; to a large extent they are not covered by the NLO calculation.

Data Analysis

The corrections for detector and hadronization effects rely mostly on the MEPS model (program LEPTO) [83], where the hard scattering at the photon vertex is calculated with the LO matrix element; higher orders are taken into account with leading log DGLAP parton showers. Hadronization is performed with the Lund string model [95]. Other programs (ARIADNE [88], HERWIG [84]) are used as cross-checks. Fig. 7.3a shows R_{2+1} [234] as a function of Q^2 in comparison to the models. The data are corrected for detector effects to the hadron level. LEPTO describes this rate reasonably well. It is not understood why the other models do not describe this data at relatively large Q^2, where little theoretical ambiguity is expected.

The QCD calculations in fixed-order perturbation theory were performed with the programs PROJET [241], DISJET [242], MEPJET [79, 80], and

---

[1] The parton level needs to be carefully defined; see Sect. 4.5.

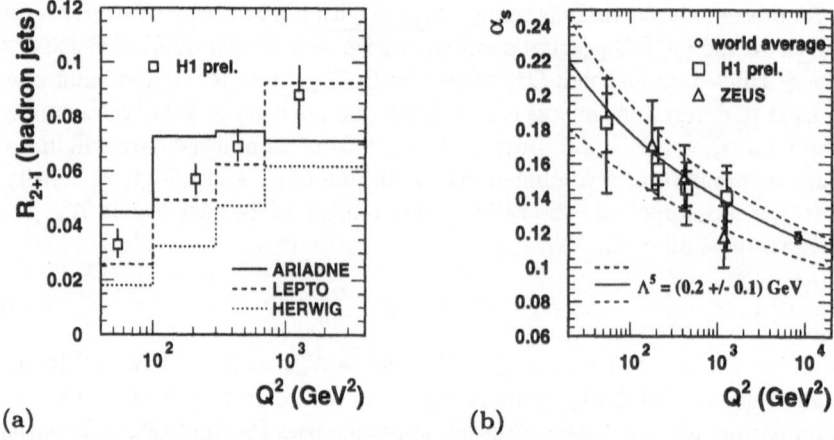

Fig. 7.3. (a) The jet rate R_{2+1} as a function of Q^2 for hadron jets (corrected for detector effects), reconstructed with the JADE algorithm with $y_{cut} = 0.02$. The preliminary H1 data [234] are compared to the models ARIADNE 4.08 [88], LEPTO 6.5 [83] and HERWIG 5.9 [84]. **(b)** The strong coupling constant α_s as a function of scale Q^2. Shown are the HERA data [234, 226] together with the world average for $\alpha_s(m_Z^2)$ [14]. The lines give the expectation for the running α_s with the QCD scales (for $n_f = 5$ flavours) $\Lambda_{\overline{MS}}^{(5)} = 0.2 \pm 0.1$ GeV

DISENT [78]. PROJET and DISJET are no longer in use, because parts of the phase space at small Q^2 and large W were not fully included.[2]

In the following two paragraphs, typical features encountered in the HERA α_s analyses are reported. One finds 3–20% 2+1 jet events in the total event sample [222, 230], depending strongly on y_{cut} and also on the event and jet selection. The dependence of R_{2+1} on y_{cut} for $Q^2 \gtrsim 100$ GeV^2 is well described by Monte Carlo (LEPTO 6.1) [222] and by NLO calculations (PROJET) [226]. With increasing y_{cut}, fewer structures are resolved as jets. According to the Monte Carlo (LEPTO 6.1), around 75% of matrix element 2+1 jet events are due to BGF for $Q^2 < 100$ GeV^2 (50% for $Q^2 > 100$ GeV^2), the rest is due to QCDC [222]. In LEPTO, around 50% of the 2+1 jet events originate not from the LO matrix element, but from the parton shower, that is mostly initial state radiation [225, 244]. The parton shower contribution is reduced at high Q^2 [245].

The NLO correction to the LO prediction is roughly 20%, depending on the analysis details, in particular on y_{cut} [226]. Migrations between jet classes due to hadronization effects can be large: $\approx$ 60% of 2+1 parton jet events and $\approx$ 90% of partonic 1+1 events are correctly classified after hadronization [222]. The resulting corrections for hadronization are sizeable, around 30% [234]. Similar numbers are obtained for migrations due to detector effects.

---

[2] The published ZEUS analysis [226] presented here was based on PROJET. It has been verified though that due to the cuts applied the effect on the α_s extraction is negligible (less than 2 per mil) [243].

In general, smearing effects become less severe for larger Q^2. Therefore the α_s analysis apply lower Q^2 cuts: 120 GeV^2 for ZEUS [226] and 40 GeV^2 for H1 [234].

Two Lorentz invariants can be introduced to describe the jet kinematics, for example $\hat{s} = (j_1 + j_2)^2$, the invariant mass squared of the two jets, and

$$z_j := \frac{P \cdot j}{P \cdot q} \approx \frac{1}{2}(1 - \cos \hat{\theta}_j). \tag{7.7}$$

The index j denotes the jet from a particular parton. $z_j \in [0,1]$ and $z_{j_1} = 1 - z_{j_2}$. The approximate relation with the scattering angle $\hat{\theta}_j$ in the parton–photon scattering frame (the rest frame of the two jet system, Fig. 7.4a) holds for small jet masses and small Q^2. In the experiment one defines $z := \min(z_{j_1}, z_{j_2})$ with $z \in [0, 0.5]$. z can be reconstructed from the measured jet angles θ_j and energies E_j in the laboratory frame,

$$z \approx \min_{j=j_1,j_2} \frac{E_j(1 - \cos \theta_j)}{\sum_{i=j_1,j_2} E_j(1 - \cos \theta_i)}. \tag{7.8}$$

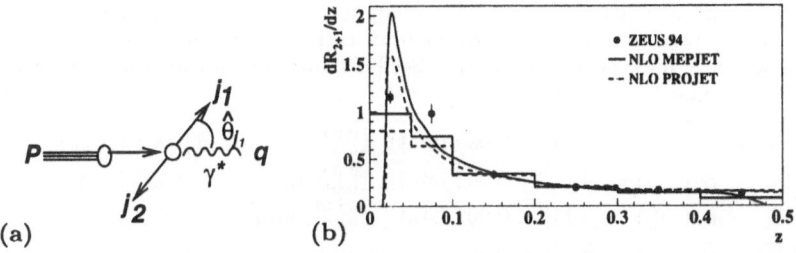

(a)   (b)

Fig. 7.4. (a) Two jet production in the centre-of-mass system of the incoming parton and the virtual photon. The scattering angle $\hat{\theta}_j$ of jet j is defined with respect to the incoming parton direction. (b) The z distribution of one of the two non-remnant jets from the ZEUS analysis [243]. The ZEUS data [226] are compared to the NLO calculations from PROJET [241] (*dashed line*) and from MEPJET [79] (*full line*). The histograms show the NLO results with the same binning as the data

The LO matrix element for the QCDC process diverges $\propto \frac{1}{(1-x/\xi)(1-z_j)}$, and for BGF $\propto \frac{1}{z_j(1-z_j)}$ with the familiar soft ($\xi \to x$) and collinear ($z_j \to 0, 1$) divergencies. It turns out that close to the collinear divergency ($z < 0.1$), more 2+1 jets are observed than expected from either the NLO calculations (see Fig. 7.4) or the LEPTO simulation [230, 226]. In this part of the phase space the NLO corrections are largest, and according to Monte Carlo, parton shower effects are most severe (apart from the LO BGF and QCDC events, QPM events with initial state parton showers can contribute up to 50% to the total 2+1 jet rate). The region $z < 0.1$, corresponding to small angles θ_{jet} in the lab system with respect to the proton remnant direction, is therefore cut out for the determination of α_s [226, 234]. H1 applies an additional cut

$\theta_{\rm jet} > 10°$, and requires $\theta_{\rm jet} < 145°$ in order to exclude the backward detector region which was not instrumented with hadronic calorimetry.

Results

From a comparison of the jet rate R_{2+1}, corrected to the parton level, with the NLO expression of 7.6, α_s is determined as a function of Q^2. ZEUS [226] has measured α_s in three Q^2 bins from 120 to 3600 GeV^2, and H1 [234, 246] in four bins from 40 to 4000 GeV^2(see Fig. 7.3). The data are consistent with the decrease of α_s with increasing scale according to the renormalization group equation. A fit to the data with the NLO expression for the running $\alpha_s(Q^2)$, (2.8), with $\alpha_s(m_Z^2)$ as free parameter yields the results in Table 7.1. Also given in Table 7.1 is the result of an H1 analysis of the 2+1 jet event rate, measured differentially in $y_{\rm cut}$ [247]. These α_s measurements from jets at HERA agree with the world average [14].

Table 7.1. α_s measurements by H1 [246, 247] and ZEUS [226] from 2+1 jet events. ZEUS quotes experimental and theoretical systematic errors separately. H1 finds that the use of alternative recombination schemes would increase the measured α_s value by the amount given (rec). For comparison, the world average from the PDG [14] is quoted. The combined result from LEP and SLC hadronic final state analyses at the Z mass is also given [218]. See [218] for a recent compilation of other α_s measurements

ZEUS [226]	$\alpha_s(m_Z^2) =$	0.117 ± 0.005 (stat) $^{+0.004}_{-0.005}$ (sys$_{\rm exp}$) ± 0.007 (sys$_{\rm th}$)
H1 [246]	$\alpha_s(m_Z^2) =$	0.117 ± 0.003 (stat) $^{+0.009}_{-0.013}$ (sys) $+0.006$ (rec)
H1 [247]	$\alpha_s(m_Z^2) =$	0.118 ± 0.002 (stat) $^{+0.007}_{-0.008}$ (sys) $^{+0.007}_{-0.006}$ (sys$_{\rm th}$)
world avg.	$\alpha_s(m_Z^2) =$	0.118 ± 0.003
e^+e^-	$\alpha_s(m_Z^2) =$	0.122 ± 0.006

For the ZEUS result, experimental and theoretical systematic errors are given separately. H1 found that alternative recombination schemes than the ones used would increase α_s by the amount quoted. The dominating systematic errors are the corrections for hadronization effects and the dependence on the choice of renormalization and factorization scales. The largest experimental systematic error is due to the uncertainty of the absolute calorimeter calibration, which is known to about $\pm 5\%$.

It is estimated [248] that with the large amount of luminosity to be collected at HERA in the future, 250 pb^{-1}, the errors on $\alpha_s(m_Z^2)$ can be reduced to ± 0.0013 (stat.) and ± 0.005 (syst.). Other algorithms than JADE may turn out to be advantageous in the future [234]. In any case, a good understanding of perturbative QCD beyond LO and hadronization will be mandatory to achieve a good precision on α_s. For example, the puzzle of why two otherwise respectable QCD models fail to describe the jet rate at the hadron level (Fig. 7.3a) needs to be resolved.

7.5 The Gluon Density

The gluon density at small fractional momentum $x_{g/p}$ $(x_{g/p} < 0.04)$ had been extracted previously only indirectly from inclusive structure function measurements. A direct measurement of the gluon density is in principle possible via the measurement of 2+1 jet events from the boson–gluon fusion (BGF) process $\gamma^* g \rightarrow q\bar{q}$. The 2+1 jet cross-section can in leading-order QCD be described as the sum of quark-initiated and gluon initiated processes, QCDC and BGF. Schematically,

$$\sigma_{2+1} = \alpha_s(Q^2) \left[A_{QCDC} \cdot q(x_{q/p}, Q^2) + A_{BGF} \cdot g(x_{g/p}, Q^2) \right]. \qquad (7.9)$$

Here $x_{q/p}$ and $x_{g/p}$ are the fractional proton momenta carried by the incoming quark and gluon. They are given in LO by $\xi = x(1 + \hat{s}/Q^2) \approx \hat{s}/W^2$, see Fig. 7.2. A_{BGF} and A_{QCDC} are coefficients that can be calculated perturbatively. The experimental challenge is to reconstruct ξ by measuring $\hat{s}$. That requires the separation in the hadronic final state of the hard subprocess from initial state parton showers and the proton remnant.

For the extraction of the gluon density $g(x_{g/p}, Q^2)$, $\alpha_s(Q^2)$ is assumed to be known and taken from the PDG world average. The contributions from the competing QCD Compton process (QCDC) $\gamma^* q \rightarrow qg$, which is calculable using known quark density functions $q(x_{q/p}, Q^2)$, and from wronlgy classified QPM events have to be subtracted statistically from the observed 2+1 jet rate.

Of particular interest is the gluon density at small $x_{g/p}$, where no direct measurements exist. In order to access small values of ξ $(= x_{g/p}$ in LO), the small invariant jet–jet mass $\sqrt{\hat{s}}$ needs to be resolved. The JADE jet algorithm resolves 2-jet events with $m_{ij}^2/W^2 > y_{cut}$. That defines the accessible phase space, and implies $\xi \approx \hat{s}/W^2 > y_{cut}$.

To reach lower values of ξ than given by the canonical $y_{cut} = 0.02$, the H1 analysis [227] uses the cone jet algorithm $(R = 1)$ applied in the hadronic CMS. For the jets $p_T > 3.5$ GeV is required. Since for massless objects

$$p_T^2 = \hat{s} z (1 - z), \qquad (7.10)$$

values of $\sqrt{\hat{s}} > 2p_T$ can be obtained. With the large $W^2 > 4400$ GeV^2, much lower ξ values can be probed than with the JADE algorithm.

From DIS events at small Q^2 $(12.5 < Q^2 < 80$ GeV$^2)$, 2+1 jet events are selected with $10° < \theta_{jet} < 150°$ in the laboratory frame. Their distance in (lab) pseudorapidity had to be smaller than $|\Delta\eta| = 2$. In the hadronic CMS, that cut translates roughly to $z > 0.1$. These angular cuts remove events affected by parton showers. $\sqrt{\hat{s}} > 10$ GeV is required. $\hat{s}$ is reconstructed (1) as the invariant mass squared of all particles belonging to the two jets, and (2) from the jet directions (pseudorapidities η_j) measured in the hadronic CMS,

$$\hat{s} = W^2 \exp(-\eta_1 - \eta_2). \qquad (7.11)$$

It is required that the two methods agree to $|\Delta\sqrt{\hat{s}}| < 10$ GeV.

The selected 2+1 jet event sample covers $0.002 < \xi < 0.2$ and $0.0003 < x < 0.0015$. According to the Monte Carlo (LEPTO), BGF and QCDC events are selected in the ratio 3:1. By unfolding the observed ξ distribution, the (LO) gluon distribution $x_{g/p} \cdot g(x_{g/p})$ at $\langle Q^2 \rangle = 30$ GeV^2 is obtained (Fig. 7.5). The data extend the region previously covered by NMC [249], $x_{g/p} > 0.04$ down to $x_{g/p} = 0.0019$. This direct measurement of the gluon density confirms the steep rise of the gluon density at small x which had been deduced from the structure function data (see Fig. 7.5). A preliminary ZEUS analysis using a similar method gives consistent results [244], within large errors. We note that sizeable NLO corrections for $2 + 1$ jet production have been calculated [80].

Recent analyses focus on the direct determination of the gluon density in NLO [250]. Using the measured partonic jet rate R_{2+1} (JADE algorithm) for $Q^2 > 40$ GeV^2 from the H1 α_s analysis [234] as input (see Sect. 7.4), the extracted gluon density for $x_{g/p} > 0.01$ is consistent within errors with results

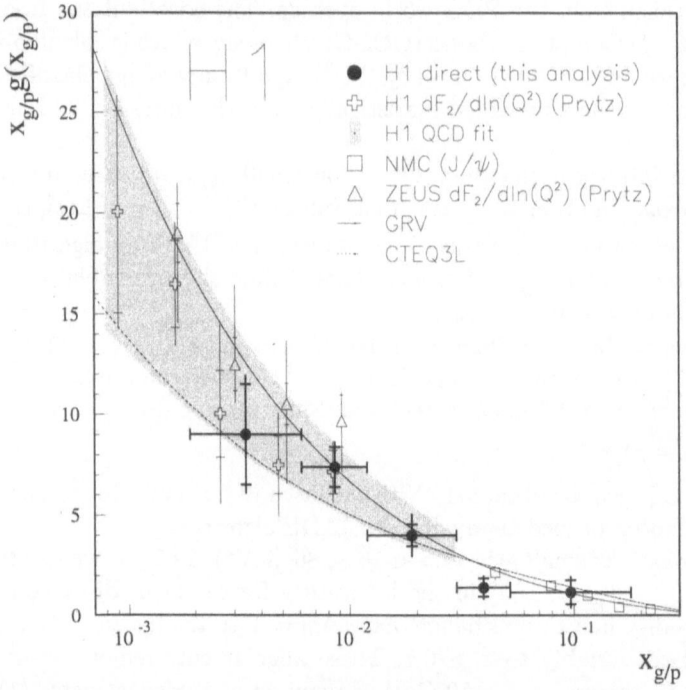

Fig. 7.5. The unfolded LO gluon distribution as a function of fractional gluon momentum $x_{g/p}$ at $\langle Q^2 \rangle$ =30 GeV^2 [227]. The direct measurement from 2+1 jet events is compared to the indirect LO determinations from structure function measurements at $\langle Q^2 \rangle$ =20 GeV^2 by H1 [253] and ZEUS [254], and to data at large x from NMC [249] obtained from J/ψ production. The other data are evolved to the $\langle Q^2 \rangle$ =30 GeV^2 of the jet analysis. Also shown are the LO gluon density parametrizations from GRV [49] and CTEQ [255]

from structure function analyses [251]. Due to the phase space restriction implied by the JADE algorithm, not very small $x_{g/p}$ are probed, $x_{g/p} > 0.01$. Smaller $x_{g/p}$ values can be accessed by tagging gluon initiated BGF events with identified charm decays [178], see Sect. 5.5.

Recent attempts to extract the NLO gluon density at small $x_{g/p}$ from dijet production at low x and Q^2 are facing a problem [231, 252]. There are about 20–30% more dijets in the data than expected from NLO calculations or from e.g. LEPTO with gluon densities as input that are compatible with the F_2 data. At face value, that would indicate a directly measured gluon density that is incompatible with the indirect extraction from the F_2 measurements. It could also be that there are effects contributing to jet production at small x and Q^2 that had not been taken properly into account. To clarify the situation, it is essential to provide data on jet production which do not contain any theoretical bias from the correction to the parton level. Such measurements are presented in the next section.

7.6 Jet Rates at Low Q^2

In a preliminary ZEUS analysis [231] that addresses the gluon density, the dijet cross section (cone algorithm in the laboratory frame with $R = 1$) has been determined and corrected for detector and hadronization effects to the parton level using LEPTO 6.3. In the laboratory frame, $\eta_{\text{jet}} < 2$ was required to exclude the forward region. A cut $p_{T\text{jet}} > 4$ GeV was applied in the laboratory frame and in the CMS.

The shapes of the measured cross-sections as a function of x, Q^2, $p_{T\text{jet}}$ and η in the CMS, and $\xi = x(1 + \hat{s}/Q^2)$ are well described by LEPTO and by NLO calculations [79, 80]. However, the absolute dijet cross-section is $\approx 34\%$ larger than that predicted by NLO QCD (see Fig. 7.6). Systematic effects cannot account for such a large discrepancy [231].

The previous jet studies were aimed at the measurement of observables defined in QCD – the strong coupling constant and the gluon density. To that end regions of phase space where the data deviate from the theory – cast either into NLO calculations or Monte Carlo generators – had been carefully avoided. These phase space regions were found at low x and low Q^2, and in the forward region close to the remnant [256, 231, 252]. The forward region at low x is discussed in Sect. 8.4. Here we shall discuss jet production at low Q^2, with Q^2 ranging from the DIS regime down to photoproduction, $0 < Q^2 < 100 \text{GeV}^2$. The object of these studies is a test of QCD and to explore the validity of QCD models for jet production. The data extend into a new phase space region in DIS, where the transverse momenta squared of the jets may exceed Q^2, $p_T^2 > Q^2$. In such a situation it may be possible to probe constituents of the virtual photon at a scale given by the p_T of the jet [257].

In a preliminary H1 analysis [229], low Q^2 DIS events are selected with $y > 0.05$, where the scattered electron angle and energy are $156° < \theta_e < 173°$

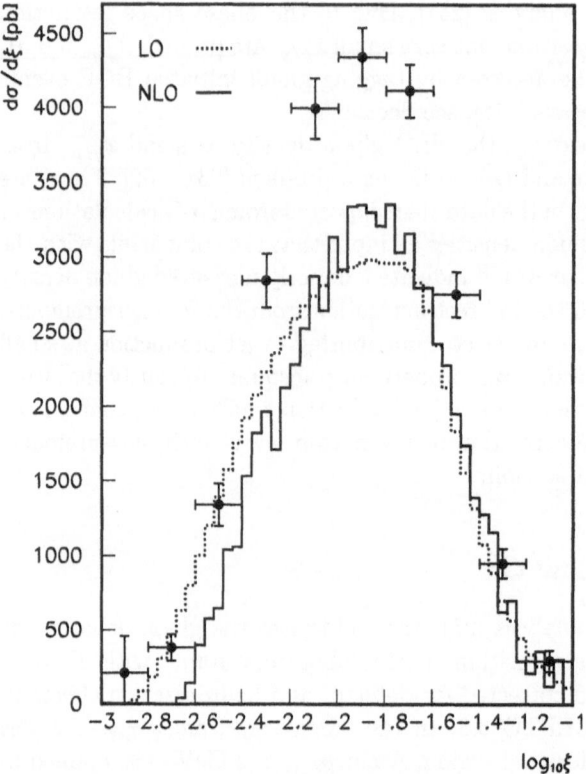

Fig. 7.6. Differential dijet cross-section (ZEUS prel. [231]) as a function of ξ, corrected to the parton level (LEPTO 6.3). Shown are statistical errors only. The data are compared to QCD calculations in LO and NLO

and $E'_e > 11$ GeV . The data sample covers $5 < Q^2 < 100$ GeV^2 and $10^{-4} < x < 10^{-2}$. Exactly two jets are required, reconstructed with the cone algorithm ($\Delta R = 1$) in the hadronic CMS and with transverse momenta $p_T > 5$ GeV . The remnant is not counted. The pseudorapidity difference $|\Delta\eta|$ between the two jets has to be less than 2. In the CMS of the scattering parton and the virtual photon, the jet angles with respect to the photon axis are given by

$$\hat{\theta}_{1,2} = 2\arctan\exp(\pm\Delta\eta/2). \tag{7.12}$$

The requirement $|\Delta\eta| < 2$ thus excludes very forward jets, because $|\cos\hat{\theta}| < 0.76$, or $z > 0.12$. Measured is the dijet rate R_2, the fraction of DIS events that fulfil the dijet requirements. The rate is corrected for detector effects to the hadron level (Fig. 7.7). The dijet rate increases from 4 to 10% for Q^2 between 10 and 100 GeV^2.

The data are compared to different QCD models, which are illustrated in see Fig. 7.8. In RAPGAP DIR (direct) [85], the virtual photon interacts directly with a proton constituent. The hardest interaction (with the highest

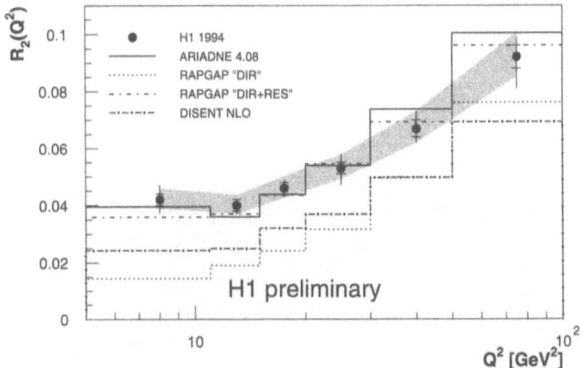

Fig. 7.7. The dijet rate R_2 as a function of Q^2 [229]. The data are corrected to the hadron level and compared to the models ARIADNE 4.08 [88] and RAPGAP 2.06 [85] with (DIR+RES) and without (DIR) a resolved component of the virtual photon. Also shown is the partonic prediction from the NLO program DISENT [78]. The shaded band shows an additional systematic error, mainly due to the uncertainty of the calorimeter calibration

p_T) takes place at the photon vertex and is modelled with the LO QCD matrix element. Softer emissions are generated with DGLAP parton showers. In RAPGAP DIR+RES it is assumed that there exists an additional contribution from so-called resolved processes, where the structure of the virtual photon is resolved by a hard scale, in this case the jet p_T with $p_T^2 > Q^2$ [258]. The hard interaction takes place further down the ladder between (evolved) partons from the photon and from the proton. Virtual photon parton densities were taken from SaS-2D [259]. The resolved contribution breaks the k_T ordering in the parton evolution.

In the colour dipole model (CDM, program ARIADNE [88]), gluon emission stems from a chain of independently radiating colour dipoles. The amount of radiation depends on the parameters that describe the extendedness of the proton remnant colour charge, and on the extendedness of the colour charge of the scattered quark, which is determined by Q^2. Emitted partons are not restricted to obeying k_T ordering. The radiation pattern thus shows similarity with BFKL evolution [89, 90].

The data cannot be described by direct interactions only (Fig. 7.7), when conventional DGLAP models are applied. At low Q^2 (and also small x), the measured dijet rate is more than a factor of two above a model which contains direct interactions only. At low Q^2 around 50% of the dijet rate is "missing"; that amounts to $\approx 2\%$ of the total event rate. Also a NLO QCD calculation for parton jets with DISENT [78] falls short of the data. Hadronization is unlikely to account for the difference, because the difference between hadronic and partonic dijet rates in the Monte Carlo was found to be < 20% for $p_T > 5$

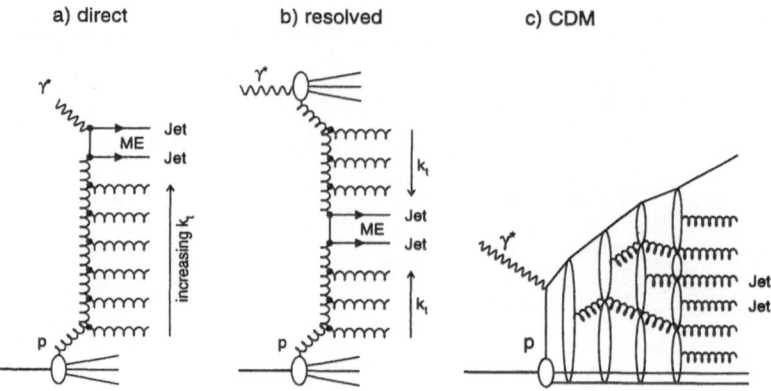

Fig. 7.8. Parton evolution scenarios. **(a)** "Direct": the hardest emission (with the largest p_T), given by the QCD matrix element, occurs at the top of the ladder. Further emissions down the ladder are ordered with decreasing k_T. **(b)** "Resolved": the hardest emission may occur anywhere in the ladder. Softer emissions decrease in k_T towards both ends. **(c)** CDM: gluons are emitted from succesively built colour dipoles. There is no restriction on k_T ordering

GeV. The discrepancy diminishes with increasing Q^2 (and also inreasing x).[3] The models where the hard interaction is not tied to the photon vertex, implementing either a resolved component of the virtual photon, or colour dipole radiation, provide a good description of the data.

In a similar analysis the single inclusive jet cross-section (k_T algorithm with $k_{cut} = 3$ GeV) has been measured as a function of jet E_T and pseudorapidity for events with $0 < Q^2 < 49$ GeV^2 and $0.3 < y < 0.6$ [228]. The data are corrected to the hadron level. It was found that when the transverse energy squared of the jet exceeds Q^2, conventional models (LEPTO 6.5 [83], HERWIG 5.9 Direct [84]) with direct photon–parton interactions alone undershoot the data (Fig. 7.9) by a large amount. A better description of the data is obtained when contributions are taken into account where the hard scale provided by the jet E_T serves to resolve partons inside the virtual photon (HERWIG 5.9 DG, with resolved photon component), or by the colour dipole model (ARIADNE 4.08 [88]).

In summary, the mechanism for jet production at small x and Q^2 is not yet very well understood. It appears that for regions of phase space where the jet p_T^2 is larger than Q^2 the conventional DIS framework, where the virtual photon resolves partons inside the proton, ceases to be valid. Instead, the high p_T of the jet could resolve both structures of the proton and of the virtual photon. We note that the linked dipole chain model [24, 91] should be able to treat both situations consistently where the jet p_T^2 is smaller or larger than Q^2. This field of research will continue with more detailed measurements. It has to be seen also in connection with "small-x" physics, to be discussed

---

[3] Recently the strong sensitivity of the comparison of the data with QCD predictions on the phase space selection for the two jets has been recognized [260].

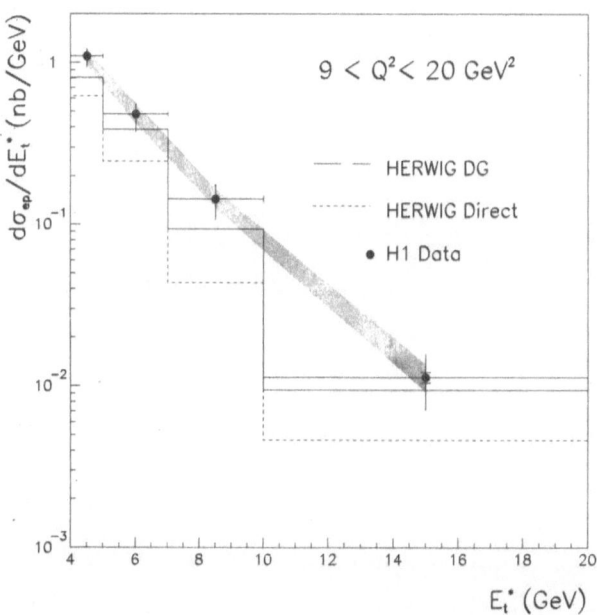

Fig. 7.9. The single inclusive jet E_T cross-section, corrected to the hadron level, for jets with CMS $2.5 < \eta^* < 0.5$ (photon direction at $+z$) [228]. The events are selected with $9 < Q^2 < 20$ GeV^2 and $0.3 < y < 0.6$. The inclusive DIS cross-section in this kinematic range is ≈ 9.8 nb . The data are compared to the HERWIG model with only direct interactions (Direct), and with resolved contributions added (DG, where the resolved photon structure is given by [261]). The shaded band shows an additional systematic error, mainly due to the uncertainty of the calorimeter calibration

in Chap. 8. Are these models alternatives to genuine "small-x" effects, as expected from BFKL evolution, or are they just a different language for the same effect? It remains to be seen whether the competing models are actually consistent with the large body of other hadronic final state data, or whether they can describe just one facet of the hadronic final state.

8. Low-x Physics

8.1 Introduction

Amongst the most interesting issues of HERA physics is QCD in the newly accessible regime of small x, see Sect. 2.5. The observed rise of the structure function F_2 towards small x suggests a strong increase of the parton density in the proton, but what is its dynamical origin? Is DGLAP evolution sufficient to account for the data, or is BFKL dynamics at work, at least to some extent? The structure function measurements are probably too inclusive to answer such questions unambiguously. Complementary measurements of the hadronic final state provide more detailed information on the reaction, which may help to uncover the underlying dynamics.

In the simple quark parton model, a quark is scattered out of the proton by the virtual boson emitted from the scattering lepton. QCD modifies this picture. Partons may be radiated before and after the boson–quark vertex, and the boson may also fuse with a gluon inside the proton by producing a quark–antiquark pair. In fact, the parton which is probed by the boson may be the end point in a whole cascade of parton branchings (see Fig. 8.1). This parton shower materializes in the hadronic final state, allowing experimental access to the dynamics governing the cascade.

Though there may be other dynamical features to be discovered that leave their footprint in the hadronic final state (see for example Sect. 7.6 on the virtual photon structure or Chap. 9 on instantons), most dedicated small x measurements have concentrated on the predicted signals for BFKL evolution. The leading log DGLAP resummation corresponds to a strong ordering of the transverse momenta k_T (w.r.t. the proton beam) in the parton cascade, see Fig. 8.1 ($Q_0^2 \ll k_{T1}^2 \ll \cdots \ll k_{Ti}^2 \ll \cdots \ll Q^2$). In the BFKL regime, the transverse momenta follow a kind of random walk ($k_{Ti}^2 \approx k_{Ti+1}^2$) [262]. The longitudinal momentum fractions are ordered according to $x_i \gg x_{i+1}$ for BFKL and $x_i > x_{i+1}$ for DGLAP.

A radiated parton (or any other object) with CMS momentum fraction x_i and longitudinal and transverse momentum components p_{zi}, p_{Ti} is to be found at CMS rapidity

$$y_i = \frac{1}{2} \ln \frac{E_i + p_{zi}}{E_i - p_{zi}} = \ln \sqrt{\frac{E_i^2 - p_{zi}^2}{(E_i - p_{zi})^2}} \approx \ln \frac{p_{Ti}}{x_i W} \approx \ln \frac{p_{Ti}}{x_i \sqrt{Q^2/x}}. \quad (8.1)$$

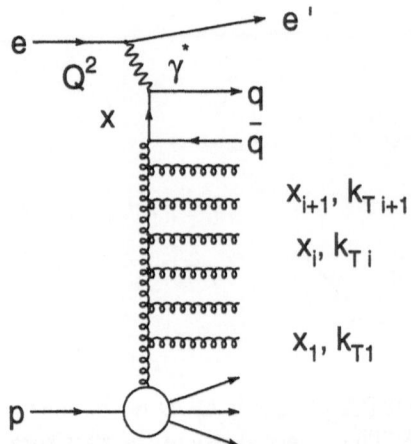

Fig. 8.1. Generic diagram for parton evolution at small x. A gluon ladder evolves between the quark box attached to the virtual photon and the proton. The gluon longitudinal momentum fractions and transverse momenta are labelled x_i and k_{Ti}

For fixed W and fixed p_{Ti}, the CMS rapidity y_i is determined by $\ln x_i$. Here the relations $E_i = x_i W/2$ and $p_{zi} \approx -x_i W/2$ in the CMS have been used (with the proton direction along the $-z$ axis). Finally,

$$x_i = \sqrt{\frac{x}{Q^2}} \cdot p_{Ti} \exp(-y_i) \qquad\qquad y_i = -\frac{1}{2} \ln \frac{x_i^2 Q^2}{x p_{Ti}^2}. \qquad (8.2)$$

For DGLAP evolution the phase space for parton radiation between the remnant and the current is restricted by k_T ordering. A generic signal for deviations from DGLAP evolution is therefore enhanced activity in the central and forward CMS rapidity region, between the current system and the proton remnant. The observables which have so far been exploited are the following:

- **Transverse energy flow:** Increased parton activity should result in increased transverse energy at central (CMS) rapidity [263].
- **Forward jets:** High energy jets with $p_{T\text{jet}}^2 \approx Q^2$ (kinematically bound to be measured in the forward calorimetric systems) could tag events with BFKL evolution, because DGLAP evolution is not allowed [264, 265].
- **Hadrons at large p_T:** High k_T partons, disfavoured by the strong k_T ordering in DGLAP, are signalled by measureable high p_T hadrons [266].
- **Other observables:** Other observables in the DIS hadronic final state like correlations and de-correlations [267, 268], multijet production [269, 270] and photon production [271] are being discussed.

Apart from a few theoretical calculations in the resummed DGLAP and BFKL schemes, and calculations in fixed-order perturbation theory (mostly NLO), predictions for the final states observables can be derived from Monte

Carlo models only. Theoretical calculations have the advantage that their theoretical input is well defined. However, most of them (an exception is presented in Sect. 8.3) neglect hadronization, making it difficult to compare them with data. On the other hand, the draw-back with Monte Carlo models is that often their complexity and their flexibility to model hadronization makes it difficult to pin down which feature of their theoretical input is actually being tested when comparing to data.

HERWIG [84] and LEPTO [83] which are based upon leading log DGLAP parton showers are used as representatives for DGLAP evolution with strong k_T ordering. Unfortunately a Monte Carlo program based upon BFKL (or CCFM, which in the limits of small and large x approaches the BFKL and DGLAP evolutions) dynamics is not yet available for ep reactions, though there exist some promising developments in that direction [272, 273, 22]. Rather, the DGLAP model predictions are contrasted with the colour dipole model ARIADNE [86, 88]. In certain aspects the CDM description of gluon emission is similar to that of the BFKL evolution [89, 90]. In particular the gluons emitted by the dipoles do not obey k_T ordering along rapidity (see Sect. 4.4).

8.2 Energy Flows

The measurements of the transverse energy flow [130, 116, 117, 118, 133, 274, 131, 275, 276, 277] have had a great influence on our understanding and modelling of the hadron production mechanism in DIS. It had been noted already in the early papers [116, 131] that some standard QCD models for the hadronic final state failed to describe the measured energy flows, in particular at low x (see also Sect. 5.1). The MEPS model (LEPTO 6.1), based upon the LO QCD matrix element with DGLAP parton showers and string fragmentation, undershoots the H1 data [116] by a factor of two in the "forward region", towards the proton remnant (Fig. 8.2). In the following years, the energy flow and its dependence on kinematic conditions has been studied in great detail.

The "forward" measurement close to the beam hole at the edge of the calorimeter acceptance at $\eta_{\text{lab}} \approx 3.5$, corresponding to $\theta_{\text{lab}} \approx 4°$, is very difficult, as particles are lost in the beam pipe, and backscatter from lower angle particles hitting dead material produce background in the forward calorimeter region [278, 279, 280]. An independent measurement is provided by ZEUS, where the instrumentation and dead material distribution in the forward region are quite different. Fortunately, the data from ZEUS [275] agree with the H1 data (see Figs. 8.3 and 8.5), confirming the "forward energy crisis" [281].

The discrepancy between data and LEPTO, considered as the reference model for hadron production, triggered further investigations, both theoretically [263, 282] and experimentally [117, 118, 275, 276], into whether low x effects, for example BFKL dynamics, could be held responsible for the

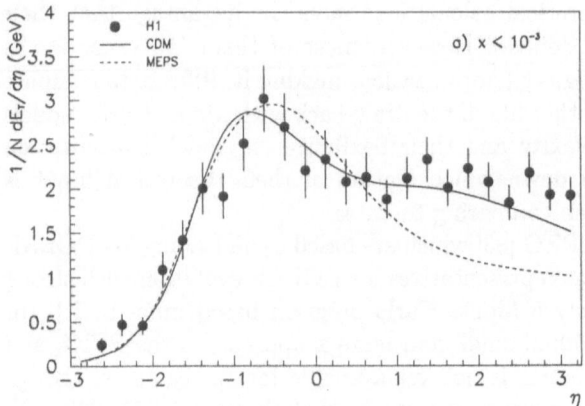

Fig. 8.2. The transverse energy flow E_T as a function of pseudorapidity η in the laboratory frame. The proton direction is to the right. The H1 data is compared to the models MEPS (LEPTO 6.1) and CDM (ARIADNE 4.03) [116]. The data are corrected for detector effects

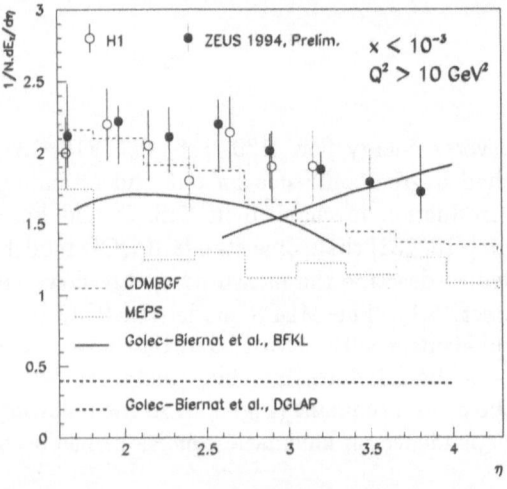

Fig. 8.3. The transverse energy flow E_T as a function of pseudorapidity η in the laboratory frame. The proton direction is to the right. The measured forward E_T flow from H1 [116] and ZEUS [275] is compared to the models MEPS (LEPTO 6.1) and CDM (ARIADNE 4.03, here labelled CDMBGF), and to BFKL and DGLAP based theoretical calculations (parton level) [263]. The data are corrected for detector effects

excess. Shown in Fig. 8.3 are also results of calculations [263] based upon either BFKL or DGLAP parton dynamics. Qualitatively, much more E_T is expected from BFKL than from DGLAP evolution, roughly at the level of the data. However, these calculations do not include hadronization, and are therefore not directly comparable to the data.

The E_T flows have now been measured over a wide range of x and Q^2 [117, 118, 276], and are presented in the hadronic CMS to eliminate the transverse boost introduced by the scattered electron. The latest preliminary H1 data [118, 283] (Fig. 8.4) also employ the plug calorimeter [284] installed at small angle behind the liquid argon calorimeter, extending the acceptance far into the target region. The plug data have large errors due to the large amount of dead material along the line of sight to the interaction vertex, between 2 and 4 nuclear absorption lengths.

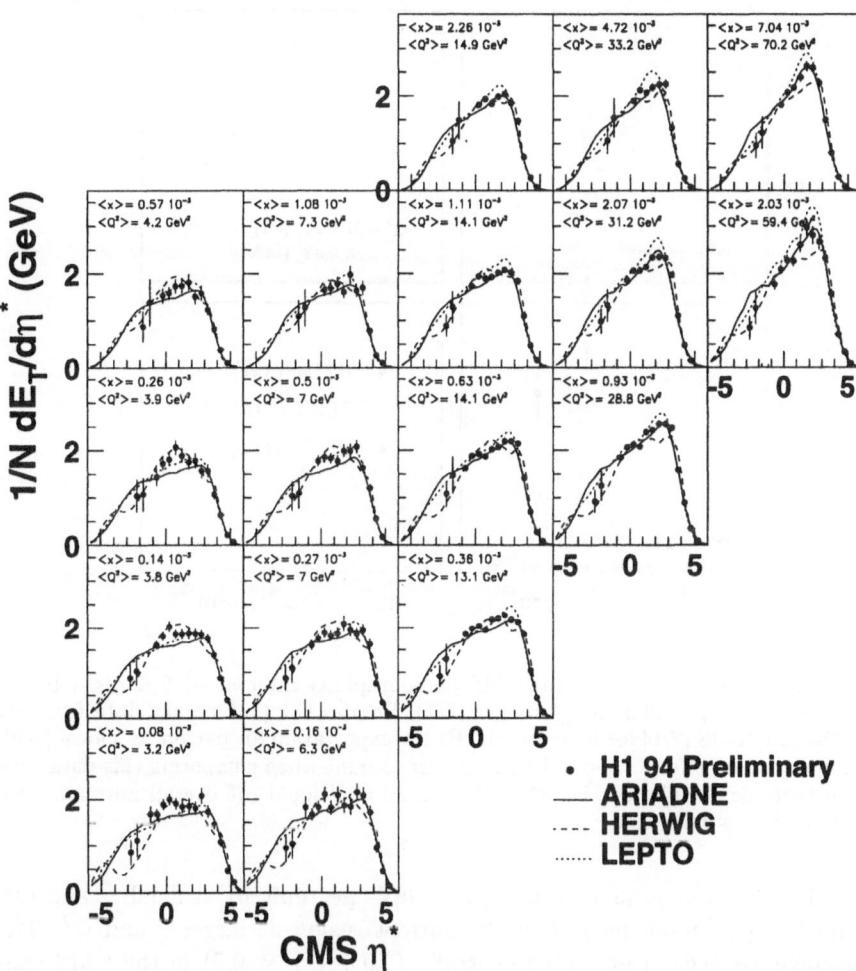

Fig. 8.4. The transverse energy flow E_T as a function of pseudorapidity η in the hadronic CMS. The proton direction is to the left. The data [118] are compared to the models CDM (ARIADNE 4.08), MEPS (LEPTO 6.4), and HERWIG 5.8. Shown are statistical errors only, except for the two foremost data points from the plug

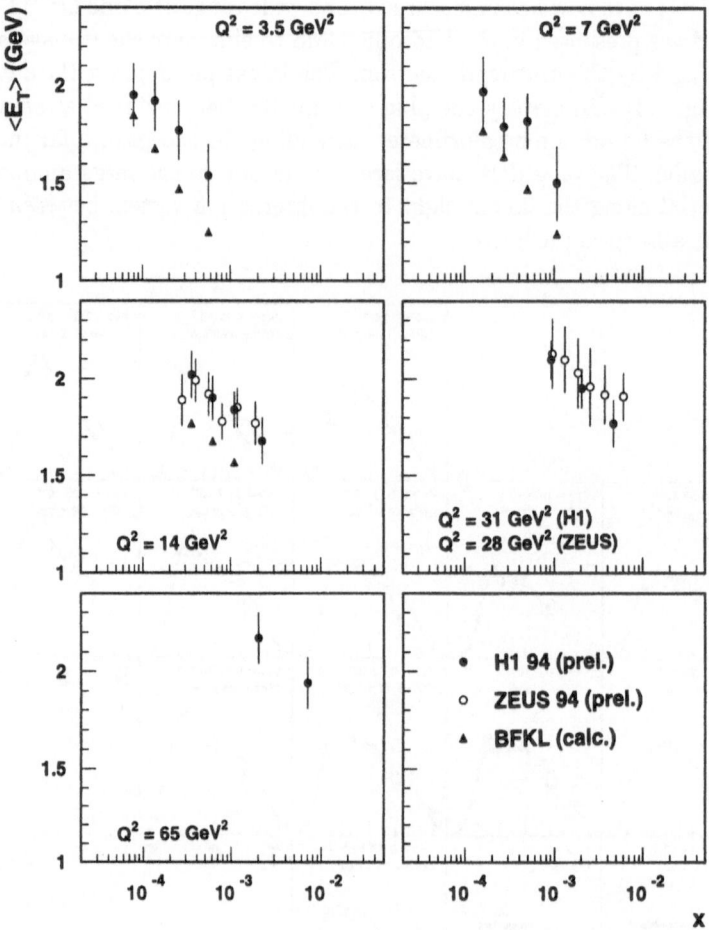

Fig. 8.5. The mean E_T in the CMS pseudorapidity interval $-0.5 < \eta < 0.5$ as a function of x for different Q^2 ranges [285]. Besides the preliminary data from H1 [118] and ZEUS [276] for hadrons, the BFKL expectation for partons is shown [263]. Hadronization effects have to be taken into account when comparing this data with the partonic calculation. For the H1 data an additional 8% overall normalization uncertainty is not shown

The E_T flow is plateau-like (≈ 2 GeV per unit η) at small x and Q^2, and becomes more peaked in the current region at larger x and Q^2. The average E_T ($\langle E_T \rangle$) measured centrally ($-0.5 < \eta < 0.5$) in the CMS rises with decreasing x (see Fig. 8.5).[1] This behaviour is predicted for the partonic

---

[1] The x dependence for fixed Q^2 implies also a W dependence. In Fig. 5.8 it is shown that the average E_T at central rapidity increases with W in DIS, and that this behaviour is in agreement with data from hadron–hadron collisions. A deeper understanding of E_T production in DIS might also be beneficial for the understanding of the hadron–hadron E_T data, which so far can only be parametrized phenomenologically.

$\langle E_T \rangle$ from the BFKL calculation, whereas for DGLAP evolution the opposite is expected [263]. CDM still gives the best overall description of the data. The DGLAP based models, MEPS and HERWIG, have developed and now give also a reasonable description of the data (Fig. 8.4).

In order to draw conclusions from comparing the data with partonic QCD calculations, hadronization effects need to be estimated from Monte Carlo models. The partons produced in the CDM give about twice as much E_T in the central rapidity region as thoses emitted from the ordered cascades in MEPS and HERWIG, see Fig. 8.6 [266], which can be traced to the fact that hard gluon radiation is much more abundant in CDM, see Fig. 8.7 [286]. However, the observable particles emerging after hadronization give rise to very similar E_T flows. The E_T flow predictions of the different models cannot be well discriminated with the present data. While hadronization adds relatively little (≈ 0.5 GeV /unit rapidity) to the partonic E_T for CDM, most of

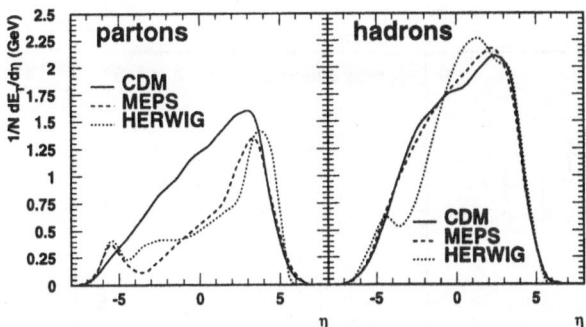

Fig. 8.6. Transverse energy flows as a function of the pseudorapidity η for partons and for hadrons. The events are generated with the models CDM (ARIADNE 4.08), MEPS (LEPTO 6.4) and HERWIG 5.8 in a "low x" kinematic bin with $\langle x \rangle = 0.00037$ and $\langle Q^2 \rangle = 13.1$ GeV^2. The proton direction is to the left

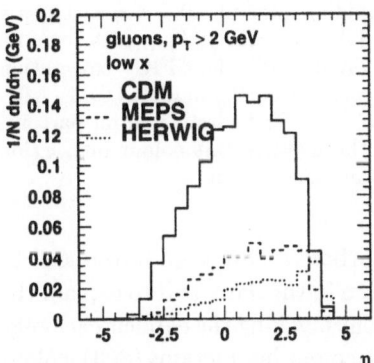

Fig. 8.7. The multiplicity of hard gluons with $p_T > 2$ GeV vs. η. The events are generated with the models CDM (ARIADNE 4.08), MEPS (LEPTO 6.4) and HERWIG 5.8 in a "low x" kinematic bin with $\langle x \rangle = 0.00037$ and $\langle Q^2 \rangle = 13.1$ GeV^2. The proton direction is to the left

the E_T (≈ 1.4 GeV /unit rapidity) is generated by hadronization for MEPS and HERWIG. It is noteworthy that CDM and MEPS use the same Lund string fragmentation model for hadronization. In addition, the partonic and the hadronic E_T are well correlated event-by-event in CDM, as opposed to the other models [266].

With the E_T data alone, the hadronization ambiguity cannot be resolved. The situation is summarized in Fig. 8.8. The $\langle E_T \rangle$ data rise with falling x and agree with both the ordered and the unordered Monte Carlo models for hadrons. BFKL also predicts a rise, as is seen in the CDM partons, but at a somewhat lower level. Apart from the normalization, the data are consistent with a BFKL interpretation, assuming hadronization effects as given by CDM. The DGLAP model partons produce much less E_T than CDM, and have the opposite x behaviour, as expected for DGLAP evolution [263]. However, the data are also consistent with DGLAP evolution, when large hadronization effects are assumed. How do they arise?

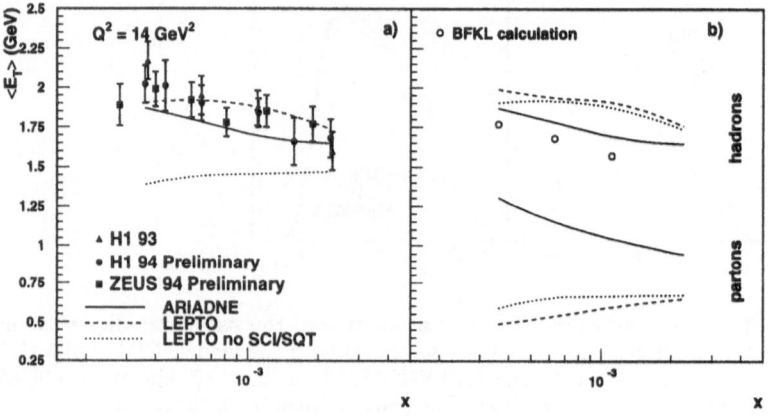

Fig. 8.8. The mean E_T in the CMS pseudorapidity interval $-0.5 < \eta < 0.5$ as a function of x for $Q^2 = 14$GeV^2. Besides the data from H1 [118] (which have an additional 8% normalization uncertainty) and ZEUS [276], the BFKL expectation for partons is shown [263], together with the results from the models CDM (ARIADNE 4.08), MEPS (LEPTO 6.4), and HERWIG 5.8 for hadrons and for partons. The LEPTO predictions for hadrons without the features soft colour interaction (SCI) and the new sea quark treatment (SQT) are also shown

In LEPTO 6.4 changes to the non-perturbative part were introduced to simultaneously describe the large E_T flow seen in the data, and to explain the rapidity gap events [106, 107, 104, 105] without invoking the explicit exchange of a Pomeron [111]. In the new concept of soft colour interactions (SCI), colour quantum numbers may be exchanged between the outgoing partons after the hard interaction, leading to a reconfiguration of the fragmenting strings (see Sect. 4.6). This may result in colour neutral subsystems, separated by a rapidity gap from the rest of the event after hadronization, or it may enhance the E_T flow, when strings are spanned back and forth (Fig. 4.14). Also the treatment of the remnant fragmentation has been changed for events where a sea quark is hit (new sea quark treatment – SQT). Instead of recombining the sea quark partner (for example an $\bar{s}$ quark in case a sea quark s was hit) within the remnant to form a hadron, a fragmenting string is now stretched between the hit sea quark and the remnant. At large x these effects are minor, but at small x they are significant, see Fig. 8.9. As long as the new SCI concept has not been ruled out by other data – from detailed studies of rapidity gap events, for example – it has to be taken seriously. The SCI does not only offer an explanation for the rapidity gap events and the large E_T flow, but has also been advocated [287] to explain the large production rate of heavy quarkonia in $p\bar{p}$ interactions at the Tevatron [288]. Presently the SCI model has difficulties describing the thrust of the hadronic final state in rapidity gap events, but so do the other models based upon Pomeron exchange [114]. Further data will be necessary to discriminate the Pomeron exchange and SCI models.

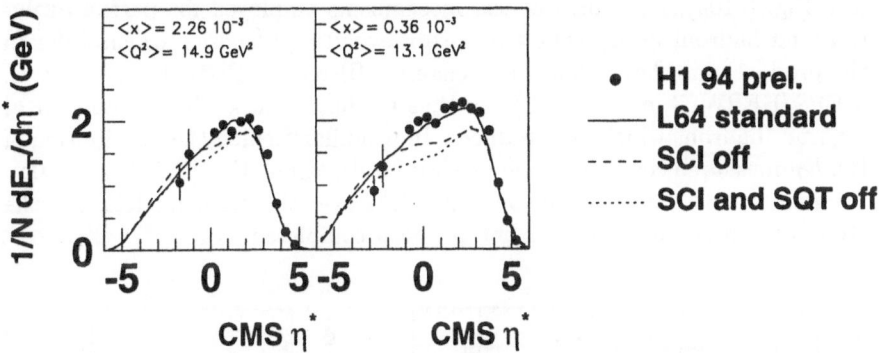

Fig. 8.9. The E_T flow from H1 [118] at $\langle x \rangle \approx 0.002$ (left) and 0.0004 (right) for $\langle Q^2 \rangle \approx 14$ GeV^2. The data (shown are statistical errors only) are compared to the MEPS model (LEPTO 6.4, structure function MRSH) in the standard version, without soft colour interactions (SCI), and with neither soft colour interactions nor the new sea quark treatment (SQT)

In the CDM hadronization effects are much smaller, because gluon radiation is more abundant, and less energy remains for hadronization. The large hadronization effects in HERWIG can possibly be explained by its very simple cluster fragmentation scheme. When there are few partons, the masses of parton pairs (clusters) to be fragmented tend to be large. In the cluster fragmentation model, they undergo a two-body decay according to phase space, leading to large transverse energies. The same mechanism leads to a significant fraction of rapidity gap events which are "not exponentially suppressed" in HERWIG without Pomeron exchange or soft colour interactions. In a later version (HERWIG 5.9) this problem is being addressed by allowing large mass clusters to split longitudinally before they decay.

8.3 Transverse Momentum Spectra

The not yet well understood hadronization effects, however interesting they may be, precluded strong conclusions on the underlying parton dynamics from the transverse energy flow measurements. It has been shown that single particle transverse momentum (p_T) spectra represent a more direct measure of the partonic activity, and corresponding measurements have been suggested [289, 266].

To answer the question of whether the E_T observed in the data is generated predominantly by parton radiation or by hadronization, inclusive p_T spectra are considered. Hadronization should produce typical spectra which are limited in p_T, while parton radiation should manifest itself in a hard tail of the p_T distribution. To test this idea particles from a "central" (in the CMS) η interval $0 < \eta < 2$ are examined. Simulated events are compared which have similar hadronic E_T (E_T^{had}) in that interval, with either a large or small contribution E_T^{par} from partons. The parton-dominated events indeed exhibit a harder p_T spectrum than the hadronization-dominated ones (see Fig. 8.10a), regardless of the mechanisms employed for parton evolution and hadronization. Different parton evolution scenarios which differ in the production of hard gluons (for example BFKL vs. DGLAP, or CDM vs. MEPS/HERWIG, see Fig. 8.3) can thus be discriminated by the number of high-p_T hadrons which are produced. In a similar fashion the distribution of the E_T measured centrally could be used for discrimination [266] (Fig. 8.10c). Preliminary H1 data [118] favour the CDM over the other models, but the study of systematic errors has not yet been completed.

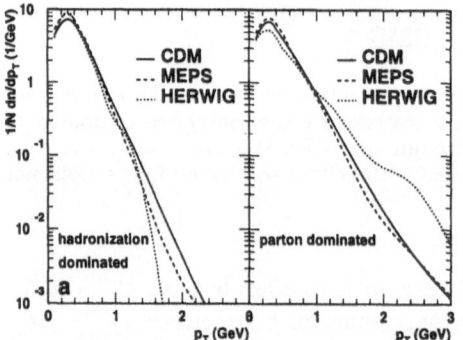
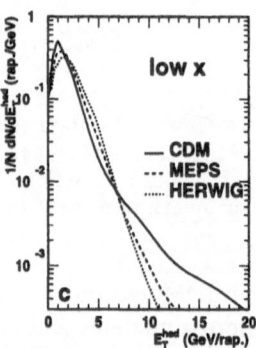

Fig. 8.10. (a) p_T spectra of charged particles with $0 < \eta < 2$ for events which are hadronization dominated (*left, E_T^{par}* < 0.2 GeV/unit rapidity) or parton dominated (*right, E_T^{par} /E_T^{had}* > 0.5) [266]. The events are required to have E_T^{had} between 1 and 2 GeV/unit rapidity. (c) The E_T distribution (measured in $0 < \eta < 2$) at $\langle x \rangle = 0.00037$ [266]. The events are generated with CDM (ARIADNE 4.08), MEPS (LEPTO 6.4) and HERWIG 5.8

H1 has measured [121] the charged particle p_T spectra as much central in the CMS ($0.5 < \eta < 1.5$) as the detector acceptance allows.[2] The following is measured

$$\frac{1}{N}\frac{dn}{dp_T} := \int_{\Delta\eta} d\eta \frac{d^4\sigma_h(ep \to e'Xh)}{d\eta dp_T dx dQ^2} \Big/ \frac{d^2\sigma(ep \to e'X)}{dx dQ^2} , \qquad (8.3)$$

where the differential cross-section to produce a hadron with transverse momentum p_T in a pseudorapidity interval $\Delta\eta$ is normalized to the differential event cross-section for given x, Q^2. In the measurement, the differential cross-sections are averaged over certain small x, Q^2 bins.

The data are compared in Fig. 8.11 to models with suppressed (LEPTO 6.4, including SCI and new SQT; HERWIG 5.8) and unsuppressed (ARIADNE 4.08) radiation patterns. At "large" x and Q^2 (here "large" means $Q^2 \approx 35\,\mathrm{GeV}^2$ and $x \approx 0.004$) all models provide a good description of the data. At smaller x and Q^2, LEPTO and HERWIG fall significantly below the data for $p_T > 1\,\mathrm{GeV}$. ARIADNE gives a good description of the data over the full kinematic range. The shortfall of the models with suppressed gluon radiation indicates that at small x there is more high k_T parton radiation present than is produced by the models based upon leading log DGLAP parton showers. These models are closer to the data in the current region, or when all charged particles are considered, including the soft ones [121]; see Fig. 5.7.

In a preliminary analysis of the H1 forward tracker, p_T data have been obtained also in the "truly central region", $-0.5 < \eta < 0.5$. In Fig. 8.12 the charged particle p_T spectra are shown for three different η ranges. At central rapidity, $-0.5 < \eta < 0.5$, the spectra from photoproduction and DIS are similar. Towards the photon fragmentation region, the DIS spectra become harder than the photoproduction data. Apparently, in that region, the virtuality Q^2 influences the hardness of the p_T spectrum, while there is little influence in the central region. A similar conlusion has been arrived at from the study of the E_T flow; see Sect. 5.1. The DIS data are well described by the CDM. LEPTO becomes too soft away from the current region. The prediction by the photoproduction model PHOJET [290] describes the photoproduction spectra well.

Theoretical calculations for the charged particle p_T spectra have become available [292], where a cross-section σ_j to produce a parton j is folded with an experimentally known fragmentation function $D_{h/j}(z, \mu^2)$ [213] to produce a hadron h with momentum fraction z from the parton j, see Fig. 8.13a. Monte Carlo models for hadronization are thus avoided. Symbolically,

$$\sigma_h = \sigma_j \otimes D_{h/j}. \qquad (8.4)$$

The parton cross-section σ_j is divided into a conventional part where gluon emissions in the ladder are ordered, and a part with unordered emis-

---

[2] The measurement relies mostly on the central tracking device; when systematic effects presently hampering the forward tracker are mastered, the measurable rapidity region could be enlarged by one unit of rapidity towards the remnant.

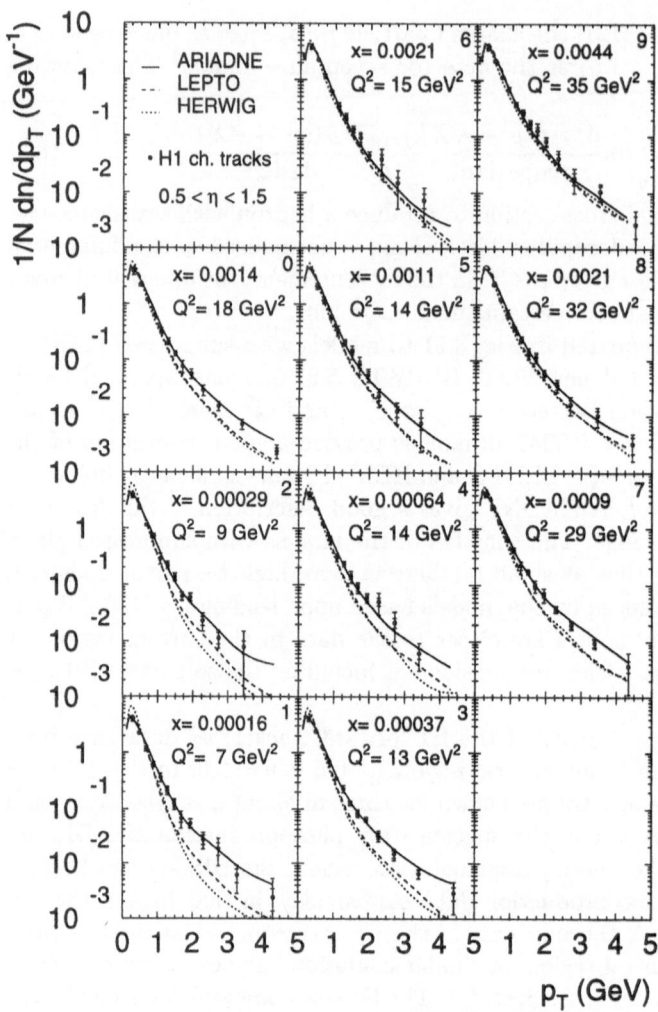

Fig. 8.11. The p_T spectra of charged particles, measured in the CMS in the pseudorapidity interval $0.5 < \eta < 1.5$ [121]. Data are shown for nine different kinematic bins with the mean values of x and Q^2 as indicated, plus the combined sample (bin 0). The models ARIADNE 4.08, LEPTO 6.4 and HERWIG 5.8 are overlayed

sions, which is calculated with the BFKL equation (see Fig 8.13a). The H1 p_T data at low x are well described by this calculation, see Fig 8.13b. The absolute normalization for the BFKL part is obtained by requiring that the calculated parton cross-section matches the measured H1 forward jets [293] at the hadron level; see Sect. 8.4. We note that this normalization neglects the effects of the jet algorithm and hadronization effects, both of which are not included in the single-parton cross-section calculation. Without the BFKL part the calculation is by a factor ≈ 2 below the data. Both predictions show little

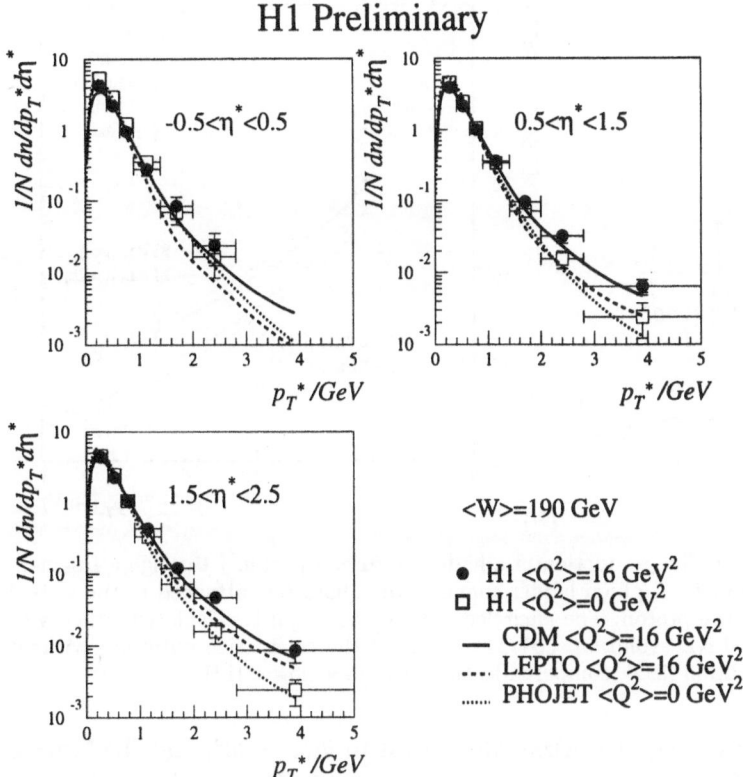

Fig. 8.12. The p_T spectra of charged particles for different CMS η ranges (H1 preliminary, [291]). DIS data with $\langle Q^2 \rangle = 16$ GeV^2 are compared with photoproduction data with $\langle Q^2 \rangle = 0$. Also shown are the DIS models CDM (ARIADNE 4.08) and LEPTO 6.5, and the photoproduction model PHOJET 1.3. The data samples are selected such that they have the same $\langle W \rangle = 190$ GeV

dependence on the choice of the factorization scale μ^2 for the fragmentation function.

For the "no BFKL" calculation the LO cross-section for three partons was used, where the main contribution is the diagram with the $q\bar{q}$ from the quark box plus one additional gluon, see Fig. 8.13a. A complete NLO calculation is still missing. DGLAP evolution was neglected, because the kinematic conditions severely disfavour DGLAP evolution. DGLAP evolution is not possible between the parton j and the photon if $k_{Tj} \approx Q$ due to strong k_T ordering. For the curves shown in Fig. 8.13b this condition is approximately fulfilled, $Q^2/2 < k_{Tj}^2 < 2Q^2$, because a leading hadron carries on average about half of the fragmenting parton's momentum, $p_T \approx k_{Tj}/2$.

An even better signal is expected when the single-particle detection capability is extended further towards the proton remnant [286, 271]. Typically

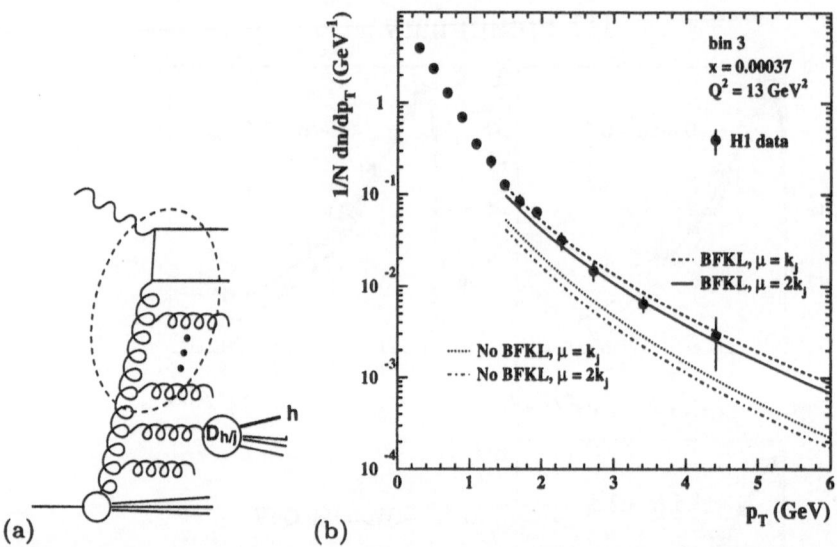

Fig. 8.13. (a) The production of a hadron h from a parton j through a fragmentation function D. A gluon ladder connects the quark box attached to the virtual photon with the proton. The encircled part of the graph is calculated either with or without BFKL evolution. (b) The result of the calculation with and without BFKL calculation [292], compared to the H1 data at low x [121]

at HERA, the central trackers are limited to $\theta_{lab} > 20°$, and the forward trackers to $\theta_{lab} > 7°$. An upgrade with a forward silicon tracker could cover angles $\theta_{lab} > 3°$ [286]. Alternatively, one can detect single photons or π^0s in the calorimeters, which have acceptance for typically $\theta_{lab} > 4°$.

H1 has measured "forward" π^0 production [294], where the π^0 meson is detected via the electromagnetic showers of their decay photons in the H1 forward liquid argon calorimeter [294]. The π^0s have been selected in the laboratory frame with $5° < \theta_{lab} < 25°$, $E > 8$ GeV , $p_T > 1$ GeV and $x_\pi := E/E_p > 0.015$ (data with other x_π cuts exist as well) in order to enhance the phase space for BFKL evolution. The number of such π^0s increases with decreasing Bjorken x; see Fig. 8.14. The models studied, MEPS and CDM, agree at large x with the data. The rise towards small x is only followed by CDM. Calculations for forward π^0 production based on [271, 292] describe well the shape of the measured x distribution. However, the normalization is found to be too large by a factor of two [295].

Coherence effects, which can be taken into account by angular ordering, are expected to modify the BFKL behaviour. They are included in the CCFM equation [20]. Predictions for the hadronic final state at low x based upon the CCFM ansatz are now emerging [22]. In general, there is less k_T diffusion in CCFM than in BFKL. The number of soft emissions is much reduced in CCFM when compared to BFKL, due to angular ordering (Fig. 8.15). For appreciable transverse momenta, the differences are not so big, affecting mainly the high multiplicity tail.

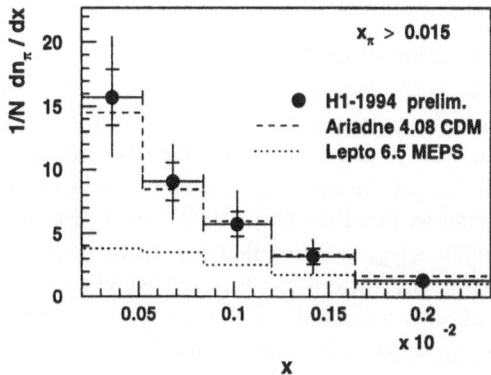

Fig. 8.14. The number of "forward" π^0 mesons as a function of Bjorken x. N is the number of events in the selected kinematic region. The preliminary H1 data [294] are compared to the MEPS model (LEPTO 6.5 [83], including SCI and the new SQT) and the CDM (ARIADNE 4.08)

The CCFM equation has been reformulated in the linked dipole chain model (LDC) [24]. First results from a Monte Carlo generator implementing the LDC in the ARIADNE framework are promising [91]. The critical distributions, E_T flows and p_T spectra at small x, are much better described than with LEPTO, though the agreement with the data is not quite as good as with the CDM [91].

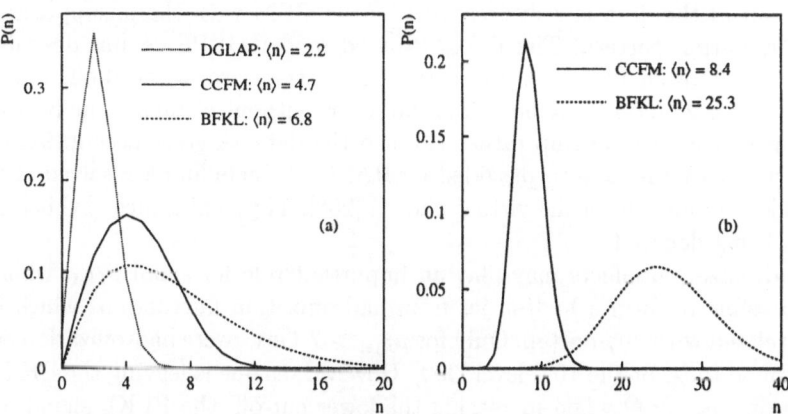

Fig. 8.15. The probability distribution $P(n)$ for the number n of gluon emissions with transverse momentum (a) $k_T > 1$ GeV and (b) $k_T > 0.007$ GeV. The calculations in the DGLAP, CCFM and BFKL schemes [22] are for events with $x = 5 \cdot 10^{-5}$

8.4 Forward Jets

The classic signature for BFKL evolution in the hadronic final state are so-called "forward jets" [264, 265] close to the remnant, that is "forward" for the HERA detectors. The selection criteria are designed such as to suppress the phase space for DGLAP evolution, and to maximize it for BFKL evolution. $x_{\text{jet}} = E_{\text{jet}}/E_p$, the ratio between the jet energy E_{jet} and the proton beam energy E_p, is required to be as large as possible, since BFKL evolution between the jet and the quark box requires $x_{\text{jet}} \gg x$ to allow for strong ordering $x_{i+1} \gg x_i$ along the ladder (see Fig. 8.1, where the forward jet is thought to emerge from a radiated gluon close to remnant). The transverse momentum $k_{T\text{jet}}$ has to be close to Q ($k_{T\text{jet}} \approx Q$). The phase space for k_T ordered DGLAP evolution ($k_{Ti+1} \gg k_{Ti}$) is thus much reduced, whereas it is still open for unordered BFKL evolution. An enhanced rate of events with such jets is thus expected in the BFKL scheme [264]. The experimental difficulty is to detect jets close to the beam hole in the proton direction in the forward calorimeter.

Previous measurements by H1 [117] were limited in statistics, but were consistent with BFKL calculations [296]. In the preliminary analysis of the larger statistics sample from 1994 [293] forward jets with $x_{\text{jet}} > 0.035$ are selected with the cone algorithm (cone radius $R = \sqrt{\Delta\eta^2 + \Delta\phi^2} = 1$) in the HERA laboratory frame within $7° < \theta_{\text{jet}} < 20°$. The transverse jet momentum requirements are $0.5 < p_{T\text{jet}}^2/Q^2 < 2$ and a lower cut-off $p_{T\text{jet}} > 3.5$ GeV. The forward jet rate, corrected for detector effects to the hadron level, increases sharply with falling x, see Fig. 8.16. This is expected from BFKL calculations, in contrast to calculations without the BFKL ladder [297]. The behaviour of the data is well described by the CDM with the unsuppressed gluon radiation pattern. The DGLAP-based model MEPS cannot describe the rise towards small x when soft colour interactions are switched off. When soft colour interactions are activated, however, without changing the parton dynamics, MEPS comes up rather close to the data. A good description of the forward jet rate is also provided by RAPGAP including a resolved (not pointlike) component of the virtual photon [258]. The significance of this fact is still being debated.

Hadronization effects may play an important rôle for small $p_{T\text{jet}}$, where it is possible to form a hadron jet from hadronization fluctuations which is uncorrelated with any parton. Only for $p_{T\text{jet}} > 7$ GeV were hadronization effects found to be at the 10% level [300]. However, as one is leaving the BFKL condition $p_{T\text{jet}} \approx Q$ when increasing the lower cut-off, the BFKL signal diminishes. This could be compensated by extending the measurements further into the forward region with upgraded detectors [301]. The hadronization effect for jets depends on the E_T produced in hadronization. In the CDM, where most of the E_T is produced by parton radiation, hadronization effects are much smaller. The MEPS forward jet rate for partons agrees with an NLO calculation with the program DISENT [78] (see Fig. 8.16). The BFKL calculation for partons shows an even larger increase towards small x than the

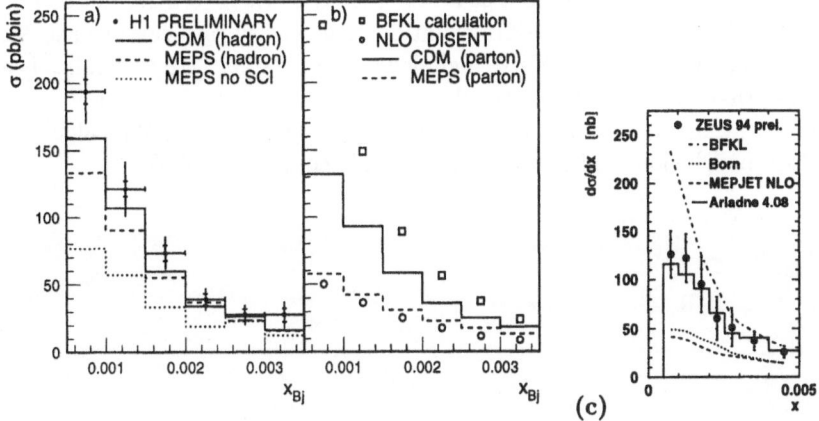

Fig. 8.16. (a) The forward jet cross-section as a function of x from H1 [293] (1994 data). The data are corrected to the hadron level and compared to the models CDM (ARIADNE 4.08) and MEPS (LEPTO 6.4 with and without soft colour interactions). (b) The forward jet cross-section from a NLO calculation [78] and for parton jets found in CDM and MEPS, compared to a BFKL parton cross-section calculation [297]. (c) The ZEUS forward jet cross-section [298] (1994 data) vs. x, corrected to the parton level [283]. $x_{jet} > 0.035$, $0.5 < p_{Tjet}^2/Q^2 < 4$ and $p_{Tjet} > 5$ GeV were required. The data are compared to an NLO jet calculation [299] and to parton jets from CDM (ARIADNE 4.08). Also shown are parton cross-sections (no jet algorithm) with ("BFKL") and without ("Born") BFKL evolution [297]. The systematic errors do not include uncertainties due to hadronization

CDM partonic forward jets. Similar conclusions were reached in a preliminary ZEUS analysis of 1994 data [298], see Fig. 8.16c.

First results from a larger data sample from 1995 have been obtained, allowing for a larger p_{Tjet} cut. In the ZEUS analysis [302, 303] events with $0.00045 < x < 0.045$, $y > 0.1$ and $E'_e > 10$ GeV are selected with $\langle Q^2 \rangle \approx$ 15 GeV^2. The jets are required to lie in the target region of the Breit frame and to satisfy $p_{Tjet} > 5$ GeV, $x_{jet} > 0.036$ and $0.5 < p_{Tjet}^2/Q^2 < 2$. In the laboratory frame $\theta_{jet} > 8.5°$ is required. Here we discuss the analysis with the cone algorithm (radius $R = 1$)[302]. An analysis with the k_T algorithm gives larger jet rates, but one arrives at similar conclusions. In a first step the data are corrected for detector effects. At the hadron level (Fig. 8.17a), the increase towards small x of the jet rate is well described by ARIADNE 4.08. LEPTO 6.5 falls short of the data by a large amount at small x.

In a second step, the data are corrected for hadronization effects to the parton level with ARIADNE, see Fig. 8.17b. Uncertainties due to this correction have not yet been studied. The data follow the expectation for jets formed from the ARIADNE partons. The partonic jet rate at low x is far above an NLO calculation. The data are also compared to a calculation with and without BFKL evolution, though these calculations do not include a jet algorithm. However, the NLO result is close to the Born graph calculation, which can be interpreted as a hint that the differences are not so large. The calculation without BFKL evolution yields a low rate close to the NLO calculation. With BFKL effects included, a much larger rate at small x is expected, that even overshoots the data. However, in view of the possibly large hadronization uncertainty and the missing jet algorithm in the calculation, no firm conclusion could be drawn from this comparison [302].

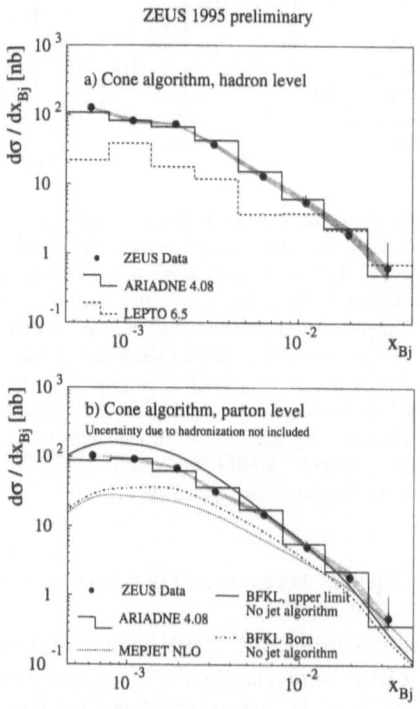

Fig. 8.17. The forward jet cross-section as a function of x from ZEUS [302]. Statistical and systematic errors are added in quadrature. The shaded bands give the additional systematic error due to the uncertainty of the hadronic energy scale. **(a)** The data are corrected to the hadron level, and compared to the expectations from ARIADNE and LEPTO. **(b)** The data are corrected to the parton level. Systematic errors due to the hadronization correction are not included. Also shown are the ARIADNE and the NLO predictions for jets [79, 80], and the BFKL and Born graph calculations for parton cross-sections [297]

8.5 Summary and Outlook

The hadronic final state has been searched for an unsuppressed parton radiation pattern, which could for example be expected from BFKL contributions, in contrast to pure DGLAP evolution. The high level of transverse energy is consistently described when BFKL evolution is included, but possibly large hadronization effects could also mimic the effect. In order to describe the data with DGLAP evolution, hadronization phenomenology has introduced new concepts like soft colour interactions in LEPTO, or it involuntarily produces "not exponentially suppressed rapidity gaps" by cluster fragmentation in HERWIG.

The forward jet measurements at low p_T are in rough agreement with BFKL expectations, but are also plagued by hadronization uncertainties. With larger statistics data samples, the measurements could be made at higher p_T where a closer relation between hadron jets and partons is expected [300]. From the measured E_T flows and forward jets, a suppressed radiation scenario is presently only tenable when large hadronization effects are assumed.

The measured single-particle p_T spectra require more parton radiation than is expected from pure DGLAP evolution, regardless of the hadronization model. The spectra have been calculated via known fragmentation functions in the DGLAP and the BFKL scheme. In the DGLAP scenario the calculation by far underestimates the data.

Whereas pure DGLAP evolution fails, BFKL evolution offers a plausible explanation of the measurements. This does not of course preclude other explanations. For example, it has been suggested that contributions from resolved photons, where the hard interaction happens not at the photon vertex but further down the ladder, could be responsible for the additional parton activity [258, 257]. We have seen possible signs for such effects also in the inclusive jet rates, see Sect. 7.6. More refined measurements to unravel the details of parton dynamics at low x, including correlations between jets and hadrons, have been proposed [58, 268]. It will also be important to check the consistency of the competing models by confronting them with the entire data on the hadronic final state.

BFKL effects are also being looked for in $p\bar{p}$ collisions at the Tevatron. A jet–jet de-correlation is seen with increasing rapidity difference, though the effect is larger than expected from BFKL, and is described by the HERWIG and PYTHIA models without BFKL dynamics [304]. Recently, BFKL effects have been brought forward to explain the discrepancy between measured jet rates and NLO calculations at small jet transverse energies for different CM energies at the Tevatron [305].

9. Instantons

9.1 Introduction

For a long time it has been recognized that the standard model contains processes which cannot be described by perturbation theory, and which violate classical conservation laws like baryon number (B) and lepton number (L) in the case of the electroweak interaction, and chirality (Q_5) in the case of the strong interaction [306]. Such anomalous processes are induced by so-called instantons [307]. The name indicates that these are non-perturbative fluctuations that are confined to "an instant" in space–time, with no corresponding free particle solutions for $t \to \pm\infty$. The interest in instantons remained somewhat academic, as observable effects were predicted to exist only at extremely high energies of $\mathcal{O}(10^5 \text{ TeV})$, until it was discovered that their exponential suppression is much reduced by the emission of gauge bosons [308]. In electroweak theory with massive gauge bosons still rather high energies of $\mathcal{O}(\gtrsim 10 \text{ TeV})$ would be required, but not so in QCD with massless gluons and strong coupling. Instanton effects could already play a rôle in QCD reactions at present day colliders. Deep inelastic ep scattering at HERA is particularly interesting, because the virtuality of the photon probe Q^2 provides a hard scale for the instanton subprocess, which is needed for theoretically sound predictions [309, 310, 311]. Instanton effects have not yet been observed in nature. Their experimental discovery would be of fundamental significance for particle physics.

Here a short introduction to the basic theoretical ideas will be given (a pedagogical treatment of instantons can be found in [312]). Instanton phenomenology in deep inelastic scattering (DIS) will be discussed, covering cross-sections and event topologies. Finally, prospects for instanton searches and first results from the analysis of HERA data will be presented.

9.2 Instanton Theory

Instantons originate from the non-trivial topological structure of the vacuum in non-abelian gauge field theories, where the vacuum is degenerate in the Chern–Simons number N_{CS}. N_{CS} is defined as an integral over the gauge fields A_μ^a with coupling g,

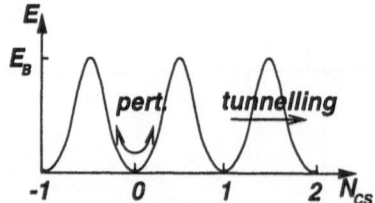

Fig. 9.1. The structure of the vacuum. Instanton solutions represent tunnelling transitions between topological inequivalent minima, which cannot be reached perturbatively

$$N_{\mathrm{CS}} := \frac{g^2}{16\pi^2} \int \mathrm{d}^3 x \epsilon_{ijk} \left(A_i^a \partial_j A_k^a - \frac{g}{3} \epsilon_{abc} A_i^a A_j^b A_k^c \right). \tag{9.1}$$

Neighbouring vacua have the same (minimal) potential energy, but differ in their topological winding number N_{CS}, and are separated by a potential barrier of height E_B (Fig. 9.1). The usual perturbative expansion of the scattering amplitudes in the coupling constant α around *one* minimum (Fig. 9.1), conveniently represented by a series of Feynman graphs, does not allow for transitions between neighbouring minima. They may however occur classically when the energy E is large enough $E > E_B$, or by quantum mechanical tunnelling when $E < E_B$, corresponding to so-called instanton solutions of the classical field equations. The transition amplitude for the instanton-induced tunnelling process is exponentially suppressed $\propto \exp(-4\pi/\alpha)$, a very small number.

In the electroweak theory, the minimal barrier height is $E_B \approx m_W/\alpha_w = \mathcal{O}(10\mathrm{TeV})$. Instanton transitions between vacua separated by ΔN_{CS} (see Fig. 9.2a for an example) would violate baryon (B) and lepton numbers ($L = L_e + L_\mu + L_\tau$) according to

$$\Delta(B + L) = -2\, n_{\mathrm{gener.}} \cdot \Delta N_{\mathrm{CS}}, \tag{9.2}$$

but respect

$$\Delta(B - L) = 0 \qquad\qquad \Delta L_e = \Delta L_\mu = \Delta L_\tau = \Delta B/3. \tag{9.3}$$

$n_{\mathrm{gener.}} = 3$ is the number of fermion generations.

In instanton induced QCD reactions (see Fig. 9.2b) chirality is violated. The chirality Q_5 is the difference between the number of left- and right-handed fermions, $Q_5 = \#L - \#R$. For n_f active quark flavours, the selection rule is

$$\Delta Q_5 = 2\, n_f \cdot \Delta N_{\mathrm{CS}}. \tag{9.4}$$

The minimal barrier height is given by the hard scale of the process, e.g. $E_B = \mathcal{O}(Q)$ for DIS [310]. The exponential suppression is less severe than in the electroweak case, because $\alpha_s \gg \alpha_w$.

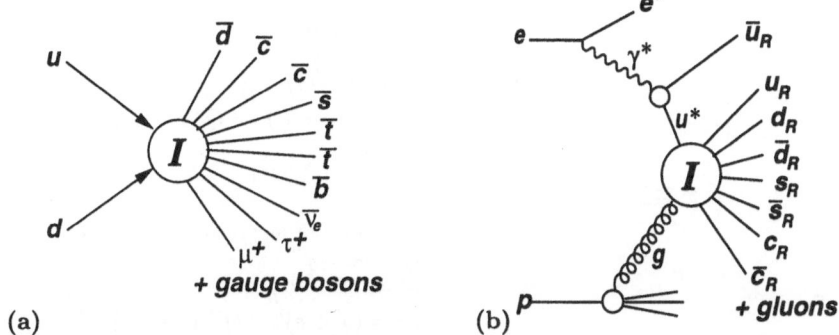

Fig. 9.2. (a) The electroweak interaction with $\Delta(B + L) = -6$ and in (b) the strong interaction with $\Delta Q_5 = 8$

9.3 Instantons at HERA

In recent years, it has been realized [310, 311, 313, 314, 315, 316] that quantitative calculations are possible for processes induced by QCD instantons in DIS due to the presence of a hard scale, Q^2. In DIS, events due to QCD instantons I (and anti-instantons $\bar{I}$) are predominantly produced in a photon-gluon fusion processes[1] (Fig. 9.3)

$$\gamma^* + g \xrightarrow{I} \sum_{n_f} (\bar{q}_R + q_R) + n_g g \qquad \gamma^* + g \xrightarrow{\bar{I}} \sum_{n_f} (\bar{q}_L + q_L) + n_g g. \qquad (9.5)$$

In each event, quarks and antiquarks of all n_f active flavours are found, with n_g gluons in addition.

The kinematics is depicted in Fig. 9.3. The DIS variables Bjorken x and Q^2 can be measured from the scattered electron, $q = e - e'$. A measurement of the other variables is more challenging. A measurement of the invariant mass of the hadronic system, excluding the remnant, would determine $\hat{s}$. If the outgoing "current jet" could be identified and measured, its 4-momentum q'' would determine $q' = q - q''$, and thus the variables x' and Q'^2 which characterise the instanton subprocess. In practice, when not all of the five independent invariants (for example $x, Q^2, x', Q'^2, \hat{s}$) can be measured, they are being integrated over.

The instanton-induced cross-section is given by a convolution of the probability to find a gluon in the proton $P_{g/p}$, the probability that the virtual photon splits into a quark–antiquark pair in the instanton background $P_{q^*/\gamma^*}^{(I)}$, and the cross-section $\sigma_{q^*g}^{(I)}(x', Q'^2)$ of the instanton subprocess [311, 313]. Multi-gluon emission enhances the cross-section [308]

---

[1] Quark-initiated processes have not yet been considered. Due to the large gluon content of the proton in the HERA domain at small x, they are expected to be of minor importance. In addition, they are expected to be suppressed by $\mathcal{O}(\alpha_s^2)$ with respect to the gluon-initiated processes.

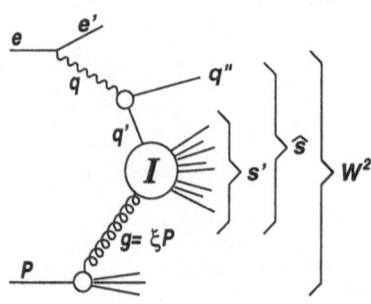

DIS variables:
$$Q^2 := -q^2$$
$$x := Q^2/(2P \cdot q)$$
$$W^2 := (q + P)^2 = Q^2(1-x)/x$$
$$\hat{s} := (q + g)^2$$
$$\xi = x(1 + \hat{s}/Q^2)$$

Variables of instanton subprocess:
$$Q'^2 := -q'^2$$
$$x' := Q'^2/(2g \cdot q')$$
$$s' := (q' + g)^2 = Q'^2(1-x')/x'$$

Fig. 9.3. Kinematics of instanton-induced processes in DIS. The labels denote the 4-vectors of the particles. A virtual photon γ^* (4-momentum $q = e - e'$) emitted from the incoming electron fuses with a gluon (4-momentum g) from the proton (4-momentum P). The gluon carries a fraction ξ of the proton momentum. The virtual quark q^* entering the instanton subprocess has 4-momentum q', and the outgoing quark from the $\gamma^* \to q\bar{q}$ splitting has 4-momentum q''. The invariant masses squared of the $\gamma^* g$ and $q^* g$ systems are $\hat{s}$ and s'. W is the invariant mass of the total hadronic system (the $\gamma^* p$ system). $0 \le x \le x/\xi \le x' \le 1$ holds. For completeness, we note $y := (Pq)/(Pe) = Q^2/(sx)$, where $s = (e + P)^2$ is the ep invariant mass squared

$$\sigma^{(I)}_{q^*g;n_g} \propto \frac{1}{n_g!}\left(\frac{1}{\alpha_s}\right)^{n_g} \exp(-4\pi/\alpha_s). \tag{9.6}$$

The cross-section of the instanton-induced subprocess is then [311]:

$$\sigma^{(I)}_{q^*g}(x', Q'^2) = \sum_{n_g=0}^{\infty} \sigma^{(I)}_{q^*g;n_g}$$

$$\approx \frac{\Sigma(x')}{Q'^2}\left(\frac{4\pi}{\alpha_s(\mu(Q'))}\right)^{\frac{21}{2}} \exp\left(\frac{-4\pi}{\alpha_s(\mu(Q'))}F(x')\right). \tag{9.7}$$

It depends critically on the functions $F(x')$ (called the "holy grail" function), which modifies the exponent in the suppression factor $\exp(-4\pi/\alpha_s)$, and on $\Sigma(x')$, which depends on $F(x')$. There exists also a scale dependence due to the choice of the renormalization scale $\mu(Q')$.

$F(x')$ can be estimated reasonably well (see Fig. 9.4) for x' not too small, $x' \gtrsim 0.2$ [311]. The extrapolation to lower values of x' is unreliable due to inherent ambiguities. In addition, multi-instanton effects should be avoided by limiting the instanton size ρ_I (the spatial region occupied during the interaction) to $\rho_I < 2$ GeV^{-1} with a cut-off $Q'^2 \gtrsim 25$ GeV^2 [311, 313]. That requirement ensures also that $\alpha_s(\mu(Q'))$ stays small enough to apply instanton perturbation theory.

The resulting instanton-induced subprocess cross-section $\sigma^{(I)}_{q^*g}(x', Q'^2)$ (see Fig. 9.5) is peaked at $Q' \approx 5$ GeV and exponentially grows with decreasing x'. The integrated instanton-induced ep DIS cross-section (see Fig. 9.6)

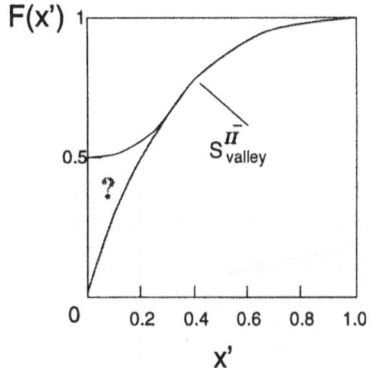

Fig. 9.4. The holy grail function $F(x')$ [311]. For small s' ($x' \approx 1$), instanton perturbation theory is applied. The calculation with the valley method matches smoothly with the perturbative result

is sizeable; for $x > 0.001$ and $x' > 0.2$ it is of $\mathcal{O}(10$ pb). The cross-section is approximately scaling (depends only on x, not on Q^2 for large Q^2) [313]. It grows towards small x, and increases dramatically when the lower x' cut-off is relaxed. Eventually higher-order instanton effects have to dampen the growth of the cross-section.

Two kinematic regions have to be distinguished. For $x' > 0.2$ the predictions are relatively safe, allowing the instanton theory to be tested. Either instantons are discovered at the predicted level – including the substantial theoretical uncertainties, which still need to be quantified – or the theory has to be revised. For $x' < 0.2$ the cross-section presumably continues to grow, but the extrapolation is extremely uncertain. For a discovery, this is the favourable region due to the large cross-section. A negative result however cannot be turned against the theory, it would rather restrict the unknown behaviour of $F(x')$ at small x'. Most promising is the kinematic region of

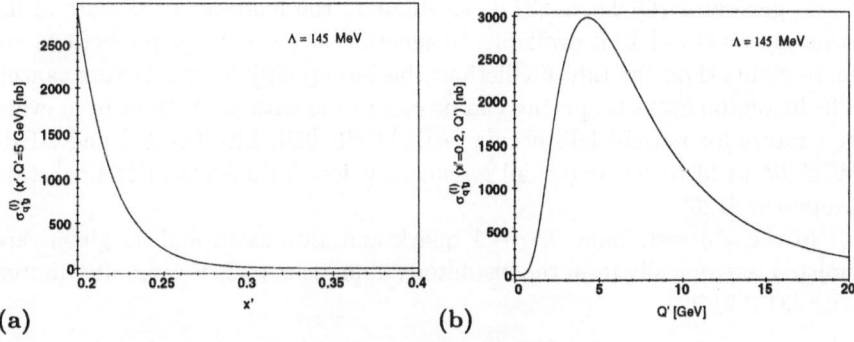

Fig. 9.5. The instanton subprocess cross-section [315] for $q^*g \rightarrow$ hadrons as a function of (a) x' for $Q' = 5$ GeV and (b) Q' for $x' = 0.2$

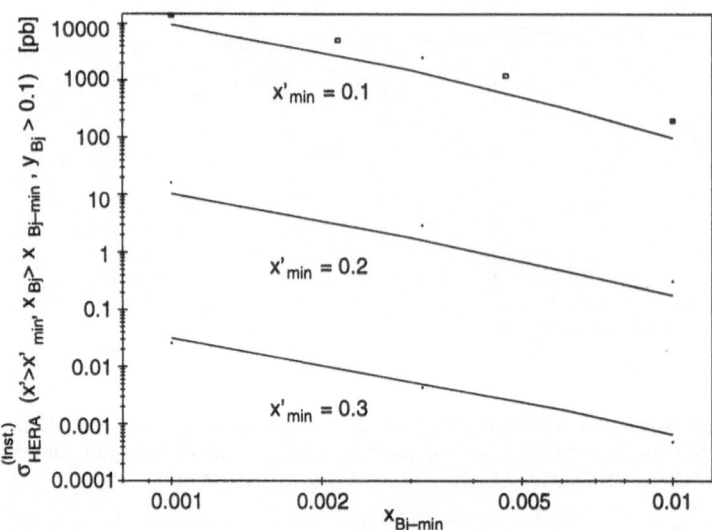

Fig. 9.6. The instanton-induced DIS cross section [315] $ep \to e'X$ integrated over Bjorken $x > x_{\min}$, $y > 0.1$, and over the regions $Q' > 5$ GeV and $x' > x'_{\min}$ as indicated

small Bjorken-x, because both the total DIS cross-section and the predicted fraction of instanton-induced events increase towards small x (see Fig. 9.9b).

9.4 Experimental Signatures

In the theoretically safe region, $x' > 0.2$, the expected fraction of instanton events in all DIS events is of $\mathcal{O}(10^{-3} - 10^{-4})$ (compare Fig.9.9b), too small to be detected in inclusive cross-section measurements (i.e. the structure function F_2). Instead, dedicated searches for the characteristic features of instanton events in the hadronic final state have to be performed. A Monte Carlo generator (QCDINS [317]) to simulate the hadronic final state of instanton events in DIS is available. In general, the event shape predictions are more stable than the rate predictions, because poorly known factors cancel. The instanton event properties can be contrasted with predictions from event generators for normal DIS events (ARIADNE [88], LEPTO [83] and HERWIG [84]) which give an overall satisfactory description of the DIS final state properties [126].

In the q^*g rest frame $2n_f - 1$ quark and antiquarks and n_g gluons are emitted isotropically from the instanton subprocess. n_g is Poisson distributed with [313, 315]

$$\langle n_g \rangle \approx \frac{2\pi}{\alpha_s} x'(1 - x') \frac{\mathrm{d}F(x')}{\mathrm{d}x'}. \tag{9.8}$$

After hadronization, this leads to a spherical system with a high multiplic-ity of hadrons, depending mainly on the available centre of mass energy $\sqrt{s'} = Q'\sqrt{1/x' - 1}$. For a typical situation ($x' = 0.2, Q' = 5$ GeV $\Rightarrow \sqrt{s'} = 10$ GeV), $\langle n_g \rangle = \mathcal{O}(2)$. About $n_p = 10$ partons and $n = 20$ hadrons are expected. The expected parton momentum spectrum is semi-hard [310] with transverse momentum $\langle p_T \rangle \approx (\pi/4)(\sqrt{s'}/\langle n_p \rangle)$.

Hadronic final state properties are conveniently being studied in the centre-of-mass system (CMS) of the incoming proton and the virtual boson, i.e. the CMS of the hadronic final state. Longitudinal and transverse quan-tities are calculated with respect to the virtual boson direction (defining the $+z$ direction). The pseudorapidity η is defined as $\eta = -\ln\tan(\theta/2)$, where θ is the angle with respect to the virtual photon direction. When boosted to the CMS, the hadrons emerging from the instanton subprocess occupy a band in pseudorapidity of half-width $\Delta\eta \approx 1$, which is homogeneously populated in azimuth [310].

The characteristics of instanton events by which they can be distinguished from normal DIS events are therefore: high multiplicity with large transverse energy; spherical event configuration (apart from the current jet); and the presence of all flavours (twice!) that are kinematically allowed in each event. One would therefore look for events which in addition to the other charac-teristics are rich in K^0s, charm decays, secondary vertices, muons, etc. In general, the strength of instanton signals in the hadronic final state increases somewhat towards low x' and large Q'^2 due to the increasing "instanton mass" $\sqrt{s'} = Q'\sqrt{1/x' - 1}$.

The "instanton band" shows up in the flow of hadronic transverse en-ergy E_T as a function of η (Fig. 9.7a). Its height and position depend on

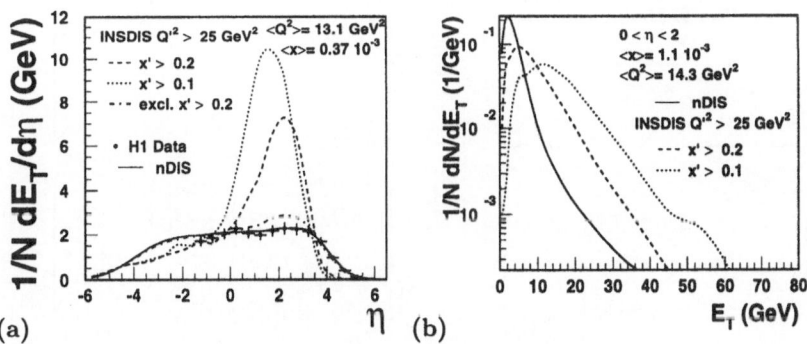

(a)   (b)

Fig. 9.7. Transverse energy E_T in the hadronic CMS [318]. **(a)** The E_T flow vs. η. The proton remnant direction is to the left. The standard QCD model (nDIS=ARIADNE) and different instanton scenarios are confronted with the H1 data [117]. The excluded scenario [318] with an instanton fraction $f_I > 11.8\%$ for $x' > 0.2$ is indicated. **(b)** The E_T distribution, where the transverse energy is measured in the CMS rapidity bin $0 < \eta < 2$, for two instanton scenarios, and the standard QCD model (nDIS=ARIADNE). The plots are normalized to the total number of events N that enter the distributions

x' and Q'^2 (and also on x and Q^2). In normal DIS events on average an E_T of 2 GeV per η unit is observed. In instanton-induced events, that number may go up to 10 GeV for low x'. A possible search strategy could involve the E_T distribution in a selected rapidity band (Fig. 9.7b), looking for high E_T events in the tail of the distribution [319, 318].

Further discrimination can be obtained [319] from the fact that for instanton events the E_T should be distributed isotropically, while normal DIS events are jet-like, in particular for large E_T. One defines

$$E_{\text{out}} := \min_i \sum_i |\boldsymbol{p}_i \cdot \boldsymbol{n}| \qquad\qquad E_{\text{in}} := \sum_i |\boldsymbol{p}_i \cdot \boldsymbol{n}'|. \qquad (9.9)$$

The sum runs over all final state hadrons i with momentum $\boldsymbol{p}_i$. $\boldsymbol{n}$ is the unit vector perpendicular to the virtual photon axis which minimizes E_{out} and thus defines the event plane. $\boldsymbol{n}'$ lies in the event plane and is normal to both $\boldsymbol{n}$ and the virtual photon axis. It is easy to show that for an ideal isotropic "instanton decay", $E_{\text{out}} = \sqrt{s'}/2$ [319]. The "instanton mass" $\sqrt{s'}$ can thus be reconstructed experimentally (Fig. 9.8a). Normal DIS events, either "1+1" or "2+1" jet events (the +1 refers to the unobserved proton remnant) are contained in the event plane, $E_{\text{out}} \ll E_{\text{in}}$, in contrast to instanton events with $E_{\text{out}} \approx E_{\text{in}}$ (see Fig. 9.8b).

Instanton events are characterized by a large particle density localized in rapidity. In normal DIS events there are about 2 charged particles per unit of pseudorapidity [121], rather uniformly distributed in η. For a low x' cut-off, that number goes up to 10 in the peak of the instanton band [318]. Very sensitive to instanton events is the charged particle multiplicity distribution [318], see Fig. 9.9a. A significant fraction of the instanton events

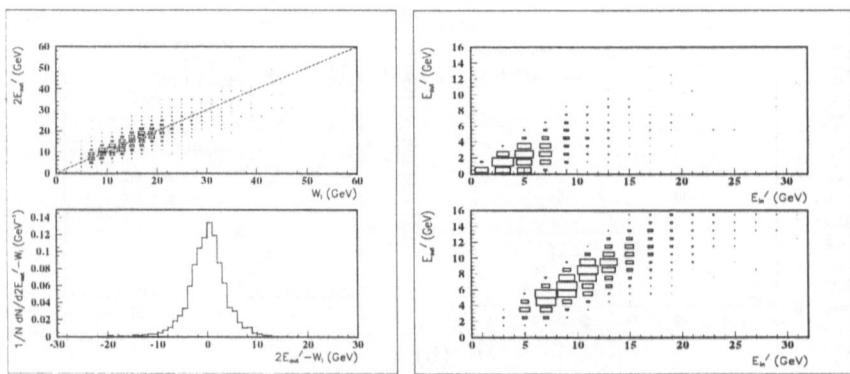

Fig. 9.8. (a) The correlation between $2 \cdot E'_{\text{out}}$ and the "instanton mass", $W_I = \sqrt{s'}$ (*top*), and the resolution for $\sqrt{s'}$ that can be achieved (*bottom*) [319]. The primes indicate additional cuts in η to minimize higher-order QCD radiation which may wash out the relation between E_{out} and $\sqrt{s'}$. **(b)** E'_{out} vs. E'_{in} for normal (*top*, HERWIG) and instanton induced events (*bottom*, QCDINS) [319]. Both distributions are taken in the hadronic CMS for events with $0.001 < x < 0.01$, $0.1 < y < 0.6$ and $20 \text{ GeV}^2 < Q^2 < 70 \text{ GeV}^2$

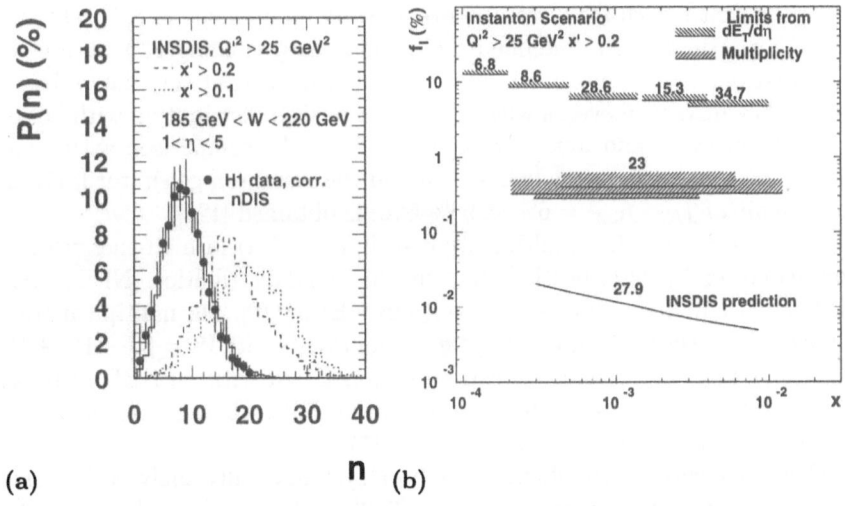

Fig. 9.9. (a) The probability distribution $P(n)$ of the charged particle multiplicity n from the CMS pseudorapidity range $1 < \eta < 5$ for events with 185 GeV $< W < 220$ GeV. Shown are the unfolded H1 data [119], the expectation from a standard DIS model (nDIS=ARIADNE)), and the predictions for instanton events with different cut-off scenarios [318]. **(b)** The maximally allowed fraction f_{lim} of instanton-induced events in DIS for $Q'^2 > 25$ GeV^2 and $x' > 0.2$ from transverse energy flows and the multiplicity distribution as a function of x [318]. Regions above the lines are excluded at 95% C.L.. The numbers give the average Q^2 values in GeV^2 for the x bins. The theory prediction, calculated with QCDINS [317], for 10 GeV$^2 < Q^2 < 80$ GeV^2 is superimposed (*full line*, label INSDIS)

would lead to charged multiplicities which are very unlikely to be found in normal DIS events. Furthermore, particle–particle correlation functions should be influenced by instanton effects [320].

9.5 Searches for Instanton Processes

The fact that instanton events look very different from the expectation for standard QCD events can be exploited to search for instanton signals in the HERA data. One strategy is to compare the shape of hadronic final state distributions to the expectation from standard QCD events (nDIS) with an admixture of instanton events (INSDIS) of fraction f_I. In case the measured distribution agrees with the standard QCD expectation, a limit on the fraction of instanton induced events in DIS $f_I < f_{\text{lim}}$ can be set. The caveat of this method is that one has to make an assumption on what standard QCD looks like. In particular at small x that issue is under debate [283, 250, 321], see Sect. 7.6 and Chap. 8. There exists a danger that an instanton effect is tuned or explained away by stretching the standard QCD predictions by generator tuning, introducing BFKL effects, etc. A good understanding of standard QCD will therefore be crucial for the positive identification of instanton effects.

In the first search for instanton events [120] an anomalous K^0 yield was looked for. For $x > 10^{-3}$ about 0.12 K^0 mesons (including $\overline{K^0}$) have been measured per event and unit pseudorapidity, with a relatively flat η distribution. For instanton events with $x' > 0.2$, a peaked distribution with about 0.55 K^0 per event and unit η is expected. From the comparison with standard QCD event generators [88, 83, 84] and the instanton generator [317] an upper limit of $f_I < f_{\text{lim}} = 6\%$ at 95% C.L. is obtained [120].

The charged particle multiplicity distribution $P(n)$ in high-energy reactions can often by described by a negative binomial distribution (NBD). Also the DIS data are relatively well described by NBDs [119]. The multiplicity distribution from the CMS interval $1 < \eta < 5$ for events with $W = 80-115$ GeV (corresponding to $x > 0.0007$) can be parametrized with an NBD of mean $\langle n \rangle = 6.90 \pm 0.33$. Possible deviations from an NBD allow for an instanton fraction of at most $f_I = 2.7\%$ at 95% C.L. [119].

Other measured event shapes have been systematically analysed in terms of their sensitivity to instanton events [318], and their dependence on the kinematic variables x, Q^2, x', Q'^2. The most sensitive distributions were the transverse energy flows [117], the pseudorapidity distribution of charged particles and their p_T spectra [121]. For example, the E_T flow has been measured over a wide range of x and Q^2, allowing us to extend the search region down to $x = 0.0001$. From a shape analysis [318] (see Fig. 9.7), instanton fractions f_I between 5 and 13% can be excluded for $x' > 0.2$ (see figs. 9.9b, 9.10). For lower x' the signal is more prominent, and somewhat better limits are obtained.

The fact that H1 [119] did not observe any events above a certain multiplicity n_{max} has been exploited [318] to place more stringent limits on instanton production.[2] A significant fraction of instanton-induced events would have multiplicities $n > n_{\text{max}}$ (compare Fig. 9.9a). Instanton fractions $f_I > 0.4-0.6\%$ can therefore be excluded for $x' > 0.2$ (see Figs. 9.9b, 9.10), and somewhat lower f_I values for a lower cut-off $x' > 0.1$ [318]. This search method has the advantage that, in contrast to the previous shape comparisons, it does not rely on assumptions for standard QCD event topologies, since no background needs to be subtracted. Unavoidable of course is the dependence on the expected instanton event shape, which may be even more uncertain than the standard QCD event shapes.

The available bounds on instanton production are summarised in Table 9.1. The most stringent limits for the theoretically "safe" scenario $x' > 0.2$ are still a factor 20 higher than what is predicted from the instanton theory, see Fig. 9.9b. Limits for other scenarios can be found in [318]. For $x' > 0.1$ they are already below the naive extrapolation into the theoretically uncertain region, providing a constraint for the theory and the holy grail function $F(x')$.

---

[2] The previous limits from the H1 multiplicity analysis [119] were derived from the shape of the multiplicity distribution for $n < n_{\text{max}}$.

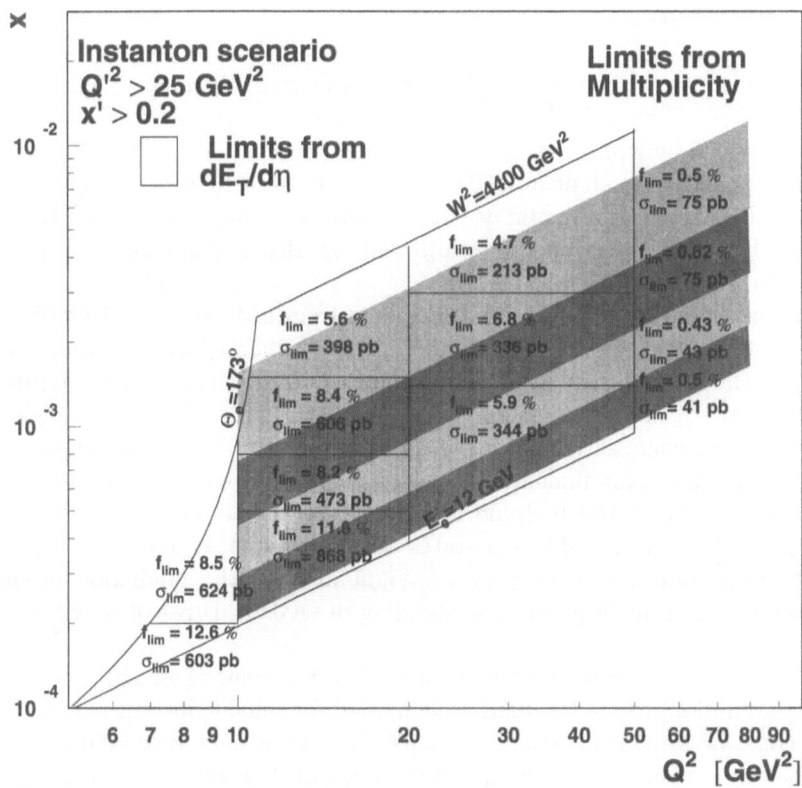

Fig. 9.10. 95% C.L. limits on instanton production with $Q'^2 > 25$ GeV^2 and $x' > 0.2$ [318]. The cross-section limits ($\sigma_{\lim}$) together with the maximally allowed instanton fraction $f_{\lim}$ are shown in the x–Q^2 plane. They are obtained from the E_T flow analysis (open fields), and from the multiplicity analysis (shaded fields) with their numbers at the right edge. The boundaries implied by the analysis cuts of the energy flow analysis in the angle and energy of the scattered electron, $\theta_e < 173°$ and $E_e > 12$ GeV, and by the requirement $W^2 > 4400$GeV^2, are indicated

Table 9.1. Limits on QCD instantons in DIS. A fraction $f_I > f_{\lim}$ of instanton-induced events in DIS is excluded at 95% C.L.

analysis	DIS kinematics covered			instanton scenario		limit
	Q^2 (GeV^2)	x	W (GeV)	Q'^2(GeV^2)	x'	$f_{\lim}$ (%)
K^0 [120]	10–70	0.001–0.01	95–230	$\gtrsim 1$	$\gtrsim 0.2$	6
multipl. [119]	10–80	0.0007–0.012	80–115	$\gtrsim 1$	$\gtrsim 0.2$	2.7
E_T flows [318]	5–50	0.0001–0.01	65–230	> 25	> 0.2	5–13
multipl. [318]	10–80	0.0001–0.01	80–220	> 25	> 0.2	0.4–0.6

9.6 Conclusion

Instanton transitions, a yet unexplored facet of non-abelian gauge field theories, have been discussed. While in the electroweak theory the $B + L$ violating effects induced by instantons are expected only at energies $\gtrsim 10$ TeV , their chirality-violating pendant in QCD could already lead to striking signatures at present day colliders. In DIS at HERA, these are a high particle multiplicity with large transverse energy localized in rapidity, and s, c and possibly b quarks in the final state.

The expected contribution to DIS events from instantons is of $\mathcal{O}(10^{-3}$–$10^{-4})$, with substantial theoretical uncertainties. First analysis of HERA data taken in the years before 1995, corresponding to an integrated luminosity of $\mathcal{O}(1.3\ \mathrm{pb}^{-1})$, are still a factor ≈ 20 above the prediction. With higher statistics data samples ($\mathcal{O}(40\ \mathrm{pb}^{-1})$ up to the end of 1997) and improved search strategies, a fundamental discovery at HERA appears to be in reach.

As the study of the hadronic final state at HERA has shown, model predictions for normal DIS final states are not unique. It will probably be safe to take some uncertainty into account also for the prediction of the instanton final state. A good understanding of QCD in DIS will be vital for instanton searches.

It might be possible to exploit also other reactions than DIS, such as photoproduction, where the hard scale needed for reliable instanton calculations could be provided by the p_T of a jet. Instantons have even be proposed [322] as an explanation of the possible excess of high Q^2 events at HERA [43, 44, 45].

10. Summary

HERA has entered a new kinematic regime in deep inelastic scattering: at small x, large W and both small and large Q^2. After five years of HERA operation, the hadronic final state has been measured in great detail. The large body of data has been confronted with QCD, the theory of the strong interaction.

We have seen progress happening at different stages:

- QCD predictions are tested experimentally, and parameters of the theory are being extracted. An example are the measurements of jet rates and the extraction of α_s or of the gluon density in the proton.
- Where the theory is less well established, there is an interplay between experimental data and theory, in which new ideas are born and developed (or rejected). Examples are soft colour interactions as an explanation for rapidity gaps and large E_T flows, or power corrections for hadronization effects.
- The data are not yet understood. Either there exist no predictions, or there are conflicting explanations. An example are the measured forward jet rates, which could be explained by BFKL evolution, or by hadronization effects, or by a resolved structure of the virtual photon.

The availability of as much data as possible over a wide kinematic range is very important for the scientific process to work efficiently. A new idea which explains one aspect of the data can be rejected early if it fails in another. The data have severely constrained QCD models for the hadronic final state which varied widely before the data had become available.

General Event Properties in Deep Inelastic Scattering at HERA

In the hadronic CMS, the produced hadrons cover about 5 units of rapidity in either hemisphere, depending mainly on the available CM energy of the hadronic system, W. There are about 2.5 charged particles per unit rapidity, and the transverse energy is on average 2 GeV per unit rapidity. Multiplicity distributions exhibit the familiar KNO scaling, and Bose–Einstein correlations are measured similar to other reactions. A large rapidity gap is found in about 10% of the events. Around 25% of the events contain charm. In the current hemisphere about 0.9 K^0 mesons and 0.15 Λ particles are measured per event. In 5–10% of the events more than one jet is found (apart from the remnant).

The global event features are largely determined by the available phase space given by the CM energy W of the hadronic system. The dependence on the photon virtuality Q^2 has also been studied. In the current region the transverse activity (E_T flow, p_T spectra, jets) increases with Q^2. In the central rapidity region Q^2 dependencies are small. Here the data are similar to what has been measured in photoproduction ($Q^2 = 0$) or hadron-hadron collisions. The particle density per unit rapidity increases more slowly than in e^+e^- annihilation with the available CM energy W. When compared with DIS data at lower W, a faster than logarithmic rise of the average multiplicity is observed.

Comparisons with QCD

Transverse energy and momentum distributions provide evidence for hard QCD radiation beyond the LO matrix element. More E_T has been measured than anticipated before HERA data taking. It is not yet clear whether soft colour interactions are required to describe the transverse energy flow, or whether not yet well established perturbative QCD effects are responsible for the large E_T measured. Soft colour interactions are of interest also because they provide a mechanism for producing rapidity gaps. Alternatively, colour dipole radiation or BFKL evolution can be employed to produce as much E_T as seen in the data.

The scaled charged particle momentum spectra measured either in the current hemispheres of the hadronic CMS with W as scale or the Breit frame with Q as scale are in general well described by QCD models in various approximations of the perturbative shower evolution in conjunction with nonperturbative hadronization models. Where applicable, perturbative QCD calculations in NLO employing fragmentation functions measured elsewhere are in agreement with the data. Scaling violations of the spectra will provide a means to measure α_s.

The growth of multiplicity with energy is described by perturbative QCD calculations in the MLLA approximation, assuming local parton–hadron duality (LPHD). The shape of the momentum spectra in the "hump-backed" form is approximately Gaussian. Its evolution with Q is consistent with the MLLA+LPHD expectation assuming soft colour coherence. Models without soft colour coherence describe the data less well. When allowance is made for the boson–gluon fusion process in DIS, the particle spectra are similar to what has been measured in e^+e^- annihilation. Evidence is found for the prediction that the soft part of the Lorentz invariant spectrum is independent of the energy of the fragmenting parton. The assumption of a constant coupling α_s is in stark conflict with the data.

Global event shape variables like thrust, jet mass and jet broadening in the Breit frame current hemisphere show an increasing event collimation with energy Q. "Power corrections" supplementing the NLO calculations provide a good description of the data. The power corrections $\propto 1/Q$ parametrize higher orders and hadronization effects and appear to be universal within 20% – the same normalization parameter can be used for different shape variables and ep as well as e^+e^- reactions. The concept of "power corrections" may turn out to be very fruitful for the understanding of hadronization effects. Assuming power corrections, the strong coupling α_s has been determined with competitive precision.

Of similar precision is the α_s determination from the rate of 2+1 jet events. The strong coupling α_s has been measured as a function of Q^2 from 2+1 jet events in a region where NLO QCD is able to describe the data, namely where Q^2 is not too small. The data are consistent with the expected Q^2 dependence of the running coupling. However, the optimal strategy for α_s measurements from jets has not yet been found. A better understanding of the connection between measured hadron jets and perturbative QCD predictions for parton jets appears to be needed. Smaller theoretical uncertainties are anticipated with other jet algorithms and increased statistics at higher Q^2 or jet p_T.

At small x and Q^2 and in particular towards the proton remnant the measured jet rates are larger than expected from either NLO calculations or conventional QCD models. "Conventional" stands for deep inelastic scattering, where a virtual photon scatters on a parton in a proton, and where the partons in the proton obey DGLAP evolution with Q^2 as scale. A good description is obtained either by the colour dipole model, or by a model in which the partonic content of the virtual photon is resolved by the large jet p_T.

A strong growth of the gluon density towards small x has been inferred from DGLAP analyses of structure function measurements. This is confirmed by direct measurements from jet and charm production, by which photon-gluon fusion processes can be tagged. The errors are still large, however. It will be very important to extend such direct measurements towards smaller x values.

Small x physics has received considerable interest. It has been speculated that the rise of F_2 with decreasing x is a sign of BFKL evolution, but DGLAP evolution was found to be able to account for the data as well. Traces of BFKL evolution have therefore been searched for in less inclusive observables in the hadronic final state, namely enhanced transverse activity in between the current and the target fragmentation regions: transverse energy flow, high p_T particles, and "forward jets". The data could not be described by conventional DGLAP evolution. Though BFKL calculations can explain the measured enhanced activity, there exist competing explanations, again making use of a virtual photon structure. It remains to be seen though whether or not such an alternative model would be in conflict with other

HERA data. In case both explanations work equally well, one might specu-
late that they have more in common than is apparent.

Finally, the possibility exists to discover QCD instanton effects at HERA.
Such a discovery would open a new field of non-perturbative QCD, and would
imply also baryon and lepton number violation in the electroweak sector
already *within* the standard model. Observable effects are predicted for the
hadronic final state, and first bounds on instanton production have been
placed. The predicted cross sections are such that it appears likely that with
increased sensitivity instantons can be discovered at HERA, or the theory in
its present form can be rejected.

References

1. A. Levy, DESY 97-013, Lectures given at "Strong Interaction Study Days", Kloster Banz, 1995
2. B. Badelek and J. Kwieciński, Rev. Mod. Phys. 68 (1996) 445
3. N. Cartiglia, hep-ph/9703245, Proc. SLAC summer school 1996;
   P. Newman, hep-ex/9707020, to appear in Proc. Workshop DIS97, Chicago 1997, eds. J. Repond and D. Krakauer;
   E. Gallo, hep-ex/9710013, Proc. Lepton-Photon Conf., Hamburg 1997
4. M. Erdmann, The Partonic Structure of the Photon, Springer Verlag 1997
5. N. Schmitz, Int. J. Mod. Phys. A3 (1988) 1997; MPI-PhE/92-23, Proc. XXII Intern. Symp. on Multiparticle Dynamics, Santiago de Compostela 1992
6. H1 Collab., I. Abt et al., Nucl. Instr. and Meth. A386 (1997) 310
7. H1 Collab., I. Abt et al., Nucl. Instr. and Meth. A386 (1997) 348
8. ZEUS Collab., M. Derrick et al., Phys. Lett. B293 (1992) 465, Z. Phys. C63 (1994) 391;
   A. Caldwell, Proc. ICHEP 92, Dallas 1992, ed. J. R. Sanford, vol. II, p. 1856
9. F. Jacquet, A. Blondel, in Proc. of the study of an ep facility for Europe, ed. U. Amaldi (1979), p. 391
10. U. Bassler and G. Bernardi, Nucl. Instr. Meth. A361 (1995) 118
11. L.N. Hand, Phys. Rev. 129 (1963) 1834
12. J. Bartels and C. Bontus, to appear in Proc. Workshop DIS97, Chicago 1997, eds. J. Repond and D. Krakauer
13. R. G. Roberts, The structure of the proton, Cambridge University Press, 1990
14. R.M. Barnett et al., Particle Data Group, Review of Particle Physics, Phys. Rev. D54 (1996) no. 1
15. Yu. L. Dokshitzer, Sov. Phys. JETP 46 (1977) 641;
    V.N. Gribov and L.N. Lipatov, Sov. J. Nucl. Phys. 15 (1972) 438 and 675;
    G. Altarelli and G. Parisi, Nucl. Phys. 126 (1977) 297
16. A. Martin, Durham preprint DTP/93/66, lectures given at XXI Intern. Meeting on Fundamental Physics, Miraflores de la Sierra 1993
17. A. DeRujula et al., Phys. Rev. D10 (1974) 1649;
    B. Badelek et al., J. Phys. G.: Nucl. Part. Phys. 22 (1996) 815
18. E.A. Kuraev, L.N. Lipatov and V.S. Fadin, Sov. Phys. JETP 45 (1972) 199;
    Y.Y. Balitsky and L.N. Lipatov, Sov. J. Nucl. Phys. 28 (1978) 282
19. M. Ciafaloni, hep-ph/9709390, Proc. Workshop "New Trends in HERA Physics", Schloß Ringberg, Tegernsee 1997, eds. B.A. Kniehl, G. Kramer and A. Wagner, p. 62
20. M. Ciafaloni, Nucl. Phys. B296 (1987) 249;
    S. Catani, F. Fiorani and G. Marchesini, Phys. Lett. B234 (1990) 339, Nucl. Phys. B336 (1990) 18
21. A.D. Martin, Proc. Workshop DIS96 on "Deep Inelastic Scattering and Related Phenomena", Rome 1996, eds. G. D'Agostini and A. Nigro, p. 156

22. G. Salam, hep-ph/9707382, Proc. Workshop "New Trends in HERA Physics", Schloß Ringberg, Tegernsee 1997, eds. B.A. Kniehl, G. Kramer and A. Wagner, p. 71;
    M. Scorletti, hep-ph/9710559, to appear in Proc. Madrid Workshop on "Low x Physics", Miraflores de la Sierra 1997
23. B. Andersson, G. Gustafson and H. Kharraziha, Phys. Rev. D57 (1998) 5543
24. B. Andersson, G. Gustafson and J. Samuelsson, Nucl. Phys. B463 (1996) 217;
    B. Andersson, G. Gustafson, H. Kharraziha and J. Samuelsson, Z. Phys. C71 (1996) 613
25. P. Landshoff, hep-ph/9410250, Proc. "Hadronic Aspects of Collider Physics", Zuoz 1994
26. P. D. B. Collins and A. D. Martin, Hadron Interactions, Adam Hilger Ltd., Bristol 1984
27. J. R. Forshaw and D. A. Ross, Quantum Chromodynamics and the Pomeron, Cambridge University Press, 1997
28. P. D. B. Collins, An Introduction to Regge Theory and High Energy Physics, Cambridge University Press, 1977
29. WA91 Collab., S. Albatzis et al., Phys. Lett. B321 (1994) 509
30. A. Donnachie and P.V. Landshoff, Phys. Lett. B296 (1992) 227
31. S. Levonian, hep-ph/9612206, Proc. ICHEP 96, Warsaw 1996, p. 17
32. M. Froissart, Phys. Rev. 123 (1961) 1053;
    A. Martin, Nuovo Cimento 42 (1966) 930
33. ZEUS Collab., M. Derrick et al., Phys. Lett. B316 (1993) 412; Z. Phys. C65 (1995) 379
34. ZEUS Collab., M. Derrick et al., Z. Phys. C69 (1996) 607; Z. Phys. C72 (1996) 399
35. ZEUS Collab., J. Breitweg et al., Phys. Lett. B407 (1997) 432
36. B. Surrow for the ZEUS Collab., to appear in Proc. Workshop DIS97, Chicago 1997, eds. J. Repond and D. Krakauer
37. H1 Collab., I. Abt et al., Nucl. Phys. B407 (1993) 515
38. H1 Collab., T. Ahmed et al., Nucl. Phys. B439 (1995) 471
39. H1 Collab., S. Aid et al., Nucl. Phys. B470 (1996) 3
40. H1 Collab., C. Adloff et al., Nucl. Phys. B497 (1997) 3
41. H1 Collab., contrib. paper 260 to HEP97, Jerusalem 1997
42. H1 Collab., C. Adloff et al., Phys. Lett. B393 (1997) 452
43. H1 Collab., C. Adloff et al., Z. Phys. C74 (1997) 191
44. ZEUS Collab., J. Breitweg et al., Z. Phys. C74 (1997) 207
45. B. Straub for the H1 and ZEUS Collab., Lepton-Photon Conference, Hamburg 1997;
    E. Elsen for the H1 and ZEUS Collab., EPS-HEP97 Conference, Jerusalem 1997
46. E665 Collab., M.R. Adams et al., Phys. Rev. D54 (1996) 3006
47. NMC Collab., M. Arneodo et al., Phys. Lett. B364 (1995) 107
48. BCDMS Collab., A.C. Benvenuti et al., Phys. Lett. B223 (1989) 485; CERN-EP/89-06
49. M. Glück, E. Reya and A. Vogt, Z. Phys. C67 (1995) 433; C53 (1992) 127; C48 (1990) 471
50. R.D. Ball and S. Forte, Phys. Lett. B335 (1994) 77, ibid. B336 (1994) 77
51. J. Kwieciński, A.D. Martin and A. M. Stasto, Phys. Rev. D56 (1997) 3991
52. K. Prytz, Phys. Lett. B311 (1993) 286
53. R. Devenish, Proc. Lepton-Photon Conf., Hamburg 1997
54. M. A. J. Botje, NIKHEF-97-028, to appear in Proc. Workshop DIS97, Chicago 1997, eds. J. Repond and D. Krakauer

55. NMC Collab., M. Arneodo et al., Phys. Lett. B309 (1993) 222
56. A.D. Martin, R.G. Roberts and W.J. Stirling, Phys. Lett. B387 (1996) 419
57. H.L. Lai et al., Phys. Rev. D55 (1997) 1280
58. K. Charchula and E.M. Levin, Proc. Workshop on "Physics at HERA", Hamburg 1991, eds. W. Buchmüller and G. Ingelman, vol. 1, p. 223
59. A.H. Mueller, J. Phys. G, Nucl. Par. Phys. 19 (1993) 1463
60. A.H. Mueller and H. Navelet, Nucl. Phys. B282 (1987) 727
61. R.D. Ball and S. Forte, Phys. Lett. B358 (1995) 365
62. M. A. J. Botje, NIKHEF-97-029, to appear in Proc. Workshop DIS97, Chicago 1997, eds. J. Repond and D. Krakauer
63. A. Zee, F. Wilczek and S.B. Treiman, Phys. Rev. D10 (1974) 2881;
    G. Altarelli and G. Martinelli, Phys. Lett. B76 (1978) 89
64. A.M. Cooper-Sarkar et al. , Z. Phys. C39 (1988) 281
65. A. M. Cooper-Sarkar, R. C. E. Devenish and M. Lancaster, Proc. Workshop on "Physics at HERA", Hamburg 1991, eds. W. Buchmüller and G. Ingelman, vol. 1, p. 155;
    M. W. Krasny, W. Placzek and H. Spiesberger, Proc. Workshop on "Physics at HERA", Hamburg 1991, eds. W. Buchmüller and G. Ingelman, vol. 1, p. 171;
    L. Bauerdick, A. Glazov and M. Klein, Proc. Workshop on "Future Physics at HERA", Hamburg 1995–1996, eds. A. De Roeck, G. Ingelman and R. Klanner, vol. 1, p. 77
66. H1 Collab., S. Aid et al., Z. Phys. C69 (1995) 27
67. ZEUS Collab., M. Derrick et al., Z. Phys. C63 (1994) 391
68. D.O. Caldwell et al., Phys. Rev. Lett. 40 (1978) 1222;
    S.I. Alekhin et al., CERN-HERA 87-01 (1987)
69. A. Donnachie and P.V. Landshoff, Z. Phys. C61 (1994) 139
70. A pedagogical presentation is in J. Collins, hep-ph/9705393
71. R.P. Feynman, Phys. Rev. Lett. 23 (1969) 1415
72. R.P. Feynman and R.D. Field, Phys. Rev. D15 (1977) 2590; Nucl. Phys. 136B (1978) 1
73. B.R. Webber, J. Phys. G17 (1991) 1579
74. T. Sjöstrand, CERN-TH/95-10, Proc. XV Brazilian National Meeting on Particles and Fields, Angra dos Reis, 1994
75. B. R. Webber, Cavendish-HEP-94/17, Proc. "Hadronic Aspects of Collider Physics", Zuoz 1994
76. Proc. HERA workshop, eds. W. Buchmüller and G. Ingelman, Hamburg (1991) Vol. 3
77. G. Altarelli and G. Martinelli, Phys. Lett. B76 (1978) 89;
    R. Peccei and R. Rückl, Nucl. Phys. B162 (1980) 125;
    A. Méndez, Nucl. Phys. B145 (1978) 199;
    C. Rumpf, G. Kramer and J. Willrodt, Z. Phys. C7 (1981) 337;
    J. G. Körner, E. Mirkes and G. A. Schuler, Int. J. Mod. Phys. A4 (1989) 1781
78. S. Catani and M. Seymour, Phys. Lett. B378 (1996) 287, Nucl. Phys. B485 (1997) 291
79. E. Mirkes and D. Zeppenfeld, Phys. Lett. B380 (1996) 105;
    E. Mirkes, hep-ph/9711224, Habilitationsschrift, Univ. Karlsruhe 1997
80. E. Mirkes and D. Zeppenfeld, Act. Phys. Polon. B27 (1996) 1393
81. D. Graudenz, DISASTER++ 1.0 manual, hep-ph/9710244
82. B. Pötter, G. Kramer, Eur. Phys. J. C1 (1998) 261
83. G. Ingelman, Proc. Workshop on "Physics at HERA", Hamburg 1991, eds. W. Buchmüller and G. Ingelman, vol. 3, p. 1366;
    G. Ingelman, A. Edin and J. Rathsman, Comp. Phys. Comm. 101 (1997) 108

84. G. Marchesini et al., Comp. Phys. Comm. 67 (1992) 465
85. H. Jung, Comp. Phys. Comm. 86 (1995) 147
86. G. Gustafson and Ulf Petterson, Nucl. Phys. B306 (1988);
    G. Gustafson, Phys. Lett. B175 (1986) 453;
    B. Andersson, G. Gustafson, L. Lönnblad and Ulf Petterson, Z. Phys. C43
    (1989) 625
87. F. Halzen and A. D. Martin, Quarks and Leptons, John Wiley & Sons, 1984
88. L. Lönnblad, Comp. Phys. Comm. 71 (1992) 15
89. L. Lönnblad, Z. Phys. C65 (1995) 285; CERN-TH/95-95;
    A. H. Mueller, Nucl. Phys. B415 (1994) 373
90. J. Rathsman, Phys. Lett. B393 (1997) 181
91. H. Kharraziha and L. Lönnblad, LU-TP 97-21, hep-ph/9709424
92. Yu. L. Dokshitzer, V. A. Khoze, A. H. Mueller and S. I. Troyan, Basics of
    Perturbative QCD, Editions Frontières, 1991
93. T. Sjöstrand, Comp. Phys. Comm. 39 (1986) 347, CERN-TH-6488-92 (1992);
    T. Sjöstrand and M. Bengtsson, Comp. Phys. Comm. 43 (1987) 367
94. B. Andersson, G. Gustafson and T. Sjöstrand, Z. Phys. C6 (1980) 235
95. X. Artru and G. Mennessier, Nucl. Phys. B70 (1974) 93;
    M. G. Bowler, Z. Phys. C11 (1981) 169;
    B. Andersson, G. Gustafson and B. Söderberg, Z. Phys. C20 (1983) 317, Nucl.
    Phys. B264 (1986) 29;
    B. Andersson, G. Gustafson, G. Ingelman and T. Sjöstrand, Phys. Rep. 97
    (1983) 33
96. JADE Collab., W. Bartel et al., Phys. Lett. B101 (1981) 129
97. R. D. Field and S. Wolfram, Nucl. Phys. B213 (1983) 65;
    B. R. Webber, Nucl. Phys. B238 (1984) 492
98. D. Amati and G. Veneziano, Phys. Lett. B83 (1979) 87;
    A. Bassetto, M. Ciafaloni and G. Marchesini, ibid. 207;
    G. Marchesini, L. Trentadue and G. Veneziano, Nucl. Phys. B181 (1980) 335
99. Ya. I. Azimov, Yu. L. Dokshitzer, V. A. Khoze and S. I. Troyan, Z. Phys. C27
    (1985) 65; C31 (1986) 21
100. W. Ochs, in Proc. XXIV Int. Symp. on Multiparticle Dynamics, Vietri-sul-
     Mare, Italy, eds. A. Giovannini, S. Lupia and R. Ugoccioni, p. 243;
     S. Lupia and W. Ochs, Phys. Lett. 143B (1984) 501
101. R. K. Ellis, W. J. Stirling and B. R. Webber, QCD and Collider Physics,
     Cambridge University Press, 1996
102. G. Curci, W. Furmanski and R. Petronzio, Nucl. Phys. B175 (1994) 473;
     E.G. Floratos, C. Kounnas and R. Lacaze, Nucl. Phys. B192 (1981) 417;
     W. Furmanski and R. Petronzio, Phys. Lett. B97 (1980) 437
103. B. Webber, Phys. Lett. 339 (1994) 148; B. Webber, Proc. Workshop DIS95
     on "Deep Inelastic Scattering and QCD", Paris 1995, eds. J.F. Laporte and
     Y. Sirois, p. 115;
     Yu. L. Dokshitzer and B. Webber, Phys. Lett. B352 (1995) 451, B404 (1997)
     321;
     M. Dasgupta and B. Webber, Eur. Phys. J. C1 (1998) 539
104. ZEUS Collab., M. Derrick et al., Phys. Lett. B315 (1993) 481
105. ZEUS Collab., M. Derrick et al., Z. Phys. C70 (1996) 391
106. H1 Collab., T. Ahmed et al., Nucl. Phys. B429 (1994) 477
107. H1 Collab., T. Ahmed et al., Phys. Lett. B348 (1995) 681
108. P. Bruni and G. Ingelman, DESY 93-187, Proc. EPS-HEP93, Marseille 1993,
     eds. J. Carr and M. Perrottet, p. 595
109. G. Ingelman and K. Janson-Prytz, Proc. Workshop on "Physics at HERA",
     Hamburg 1991, eds. W. Buchmüller and G. Ingelman, vol. 1, p. 233

110. G. Ingelman and P. Schlein, Phys. Lett. B152 (1985) 256
111. A. Edin, G. Ingelman and J. Rathsman, J. Phys. G22 (1996) 943; Phys. Lett. B366 (1996) 371; Z. Phys. C75 (1997) 57
112. W. Buchmüller and A. Hebecker, Phys. Lett. B355 (1995) 573
113. W. Buchmüller, M. F. McDermott and A. Hebecker, Phys. Lett. B410 (1997) 304;
     W. Buchmüller, DESY 97-122, to appear in Proc. Workshop DIS97, Chicago 1997, eds. J. Repond and D. Krakauer
114. H1 Collab., C. Adloff et al., Eur. Phys. J. C1 (1998) 495
115. H1 Collab., C. Adloff et al., Phys. Lett. B428 (1998) 206
116. H1 Collab., I. Abt et al., Z. Phys. C63 (1994) 377
117. H1 Collab., S. Aid et al., Phys. Lett. B356 (1995) 118
118. H1 Collab., contrib. paper pa02-073 to ICHEP'96, Warsaw 1996;
     M. F. Hess, Dissertation Univ. Hamburg 1996
119. H1 Collab., S. Aid et al., Z. Phys. C72 (1996) 573
120. H1 Collab., S. Aid et al., Nucl. Phys. B480 (1996) 3
121. H1 Collab., C. Adloff et al., Nucl. Phys. B485 (1997) 3
122. G. Gustafson and Jari Häkkinen, Z. Phys. C64 (1994) 659
123. Proc. Workshop "Physics at LEP2", CERN 96-01, eds. G. Altarelli, T. Sjöstrand and F. Zwirner
124. A. Kwiatkowski, H. Spiesberger and H.-J. Möhring, Comp. Phys. Comm. 69 (1992) 155
125. K. Charchula, G. Schuler, H. Spiesberger, Comp. Phys. Comm. 81 (1994) 381
126. N. Brook et al., Proc. Workshop on "Future Physics at HERA", Hamburg 1995–1996, eds. A. De Roeck, G. Ingelman and R. Klanner, vol. 2, p. 613
127. T.J. Bromley et al., Proc. Workshop on "Future Physics at HERA", Hamburg 1995–1996, eds. A. De Roeck, G. Ingelman and R. Klanner, vol. 2, p. 611
128. T. Carli, Proc. Workshop DIS96 on "Deep Inelastic Scattering and Related Phenomena", Rome 1996, eds. G. D'Agostini and A. Nigro, p. 415
129. L. Lönnblad, Z. Phys. C70 (1996) 107
130. H1 Collab., T. Ahmed et al., Phys. Lett. B298 (1993) 469
131. ZEUS Collab., M. Derrick et al., Z. Phys. C59 (1993) 231
132. J. D. Bjorken, SLAC summer school on "Deep Inelastic Electroproduction", SLAC-167 (1973), vol. 1, p. 1
133. H1 Collab., S. Aid et al., Phys. Lett. B358 (1995) 412
134. B.R. Webber, Phys. Lett. 143B (1984) 501 and references therein
135. A.H. Mueller, Phys. Lett B104 (1981) 161;
     B.I. Ermolaev and V.S. Fadin, JETP Lett. 33 (1981) 285
136. P. Carruthers and C.C. Shih, Int. J. Mod. Phys. A2 (1987) 1447;
     L. van Hove and A. Giovannini, Z. Phys. C30 (1986) 391; Act. Phys. Polon. B19 (1988) 917
137. A. Giovannini, Nucl. Phys. B161 (1979) 429
138. S. Carius and G. Ingelman, Phys. Lett. B252 (1990) 647;
     R. Szwed, G. Wrochna and A.K. Wroblewski, Mod. Phys. Lett. A6 (1991) 245;
     M. Gazdzicki et al., Mod. Phys. Lett. A6 (1991) 981
139. DELPHI Collab., P. Abreu et al., Z. Phys. C56 (1992) 63;
     ALEPH Collab., D. Buskulic et al., Z. Phys. C69 (1995) 1
140. Z. Koba, H.B. Nielsen and P. Olesen, Nucl. Phys. N40 (1972) 317
141. A.M. Polyakov, Sov. Phys. JETP 32 (1971) 296, ibid. 33 (1971) 850;
     S.J. Orfanidis and V. Rittenberg, Phys. Rev. D10 (1974) 2892;
     G. Cohen-Tannoudji and W. Ochs, Z. Phys. C39 (1988) 513;
     W. Ochs, Z. Phys. C23 (1984) 131
142. E.A. De Wolf, I.M. Dremin and W. Kittel, Phys. Rep. 270 (1996) 1

143. E.D. Malaza and B.R. Webber, Nucl. Phys. B267 (1986) 702, ibid. B419 (1984) 510
144. B. Andersson G. Gustafson, A. Nilsson and C. Sjogren, Z. Phys. C49 (1991) 79
145. UA1 Collab., C. Albajar et al., Nucl. Phys. B335 (1990) 261
146. E665 Collab., M.R. Adams et al., Z. Phys. C61 (1994) 179
147. EMC Collab., M. Arneodo et al., Z. Phys. C35 (1987) 335
148. A. Breakstone et al., Nuovo Cim. A102 (1989) 1199
149. A. De Roeck, PhD thesis, U. Antwerpen 1988
150. NA22 Collab., M. Adamus et al., Z. Phys. C37 (1988) 215
151. ZEUS Collab., M. Derrick et al., Z. Phys. C70 (1996) 1
152. EMC Collab., M. Arneodo et al., Z. Phys. C35 (1987) 417;
     EMC Collab., J. Ashman et al., Z. Phys. C52 (1991) 361
153. E665 Collab., M.R. Adams et al., Phys. Rev. D50 (1994) 1836
154. DELPHI Collab., P. Abreu et al., Phys. Lett. B311 (1993) 408
155. E665 Collab., M.R. Adams et al., Z. Phys. C76 (1997) 441
156. D. Graudenz, Phys. Lett. B406 (1997) 178; Proc. Workshop on "Future Physics at HERA", Hamburg 1995–1996, eds. A. De Roeck, G. Ingelman and R. Klanner, p. 533
157. J. Binnewies, B.A. Kniehl and G. Kramer, Z. Phys. C65 (1995) 471; Phys. Rev. D52 (1995) 4947
158. B. Andersson, G. Gustafson and T. Sjöstrand, Z. Phys. C9 (1981) 233.
159. I.G. Knowles et al., hep-ph/9601212
160. T. Carli for the H1 and ZEUS Collab., DESY 97-10, Proc. 16th Intern. Conf. on Physics in Collisions, Mexico City 1996, eds. H. Castilla et al., p. 415
161. R. Mohr, Diploma thesis, Univ. Hamburg 1997
162. A.K. Wróblewski, Proc. of the 25th Intern. Conf. on HEP, Singapore 1990, eds. K. K. Phua and Y. Yamaguchi, p. 125
163. ZEUS Collab., M. Derrick et al., Z. Phys. C68 (1995) 29
164. E665 Collab., M.R. Adams et al., Z. Phys. C61 (1994) 539
165. WA21 Collab., G.T. Jones et al., Z. Phys. C57 (1993) 197
166. EMC Collab., M. Arneodo et al., Phys. Lett. B150 (1985) 458
167. DELPHI Collab., P. Abreu et al., Z. Phys. C65 (1995) 587
168. D. Milstead, private communication
169. H1 Collab., C. Adloff et al., Z. Phys. C72 (1996) 593
170. ZEUS Collab., M. Derrick et al., Phys. Lett. B407 (1997) 402
171. S.J. Brodsky et al., Phys. Lett B93 (1980) 451, Nucl. Phys. B369 (1992) 519; G. Ingelman, L. Jönsson and M. Nyberg, Phys. Rev. D47 (1993) 4872
172. K. Daum, S. Riemersma, B.W. Harris, E. Laenen and J. Smith, Proc. Workshop on "Future Physics at HERA", Hamburg 1995–1996, eds. A. De Roeck, G. Ingelman and R. Klanner, vol. 1, p. 89
173. H1 Collab., S. Aid et al., Nucl. Phys. B468 (1996) 3
174. E. Laenen et al., Nucl. Phys. B392 (1993) 162, 229; Phys. Lett. B291 (1992) 325;
     S. Riemersma, J. Smith and W.L. van Neerven, Phys. Lett. B347 (1995) 143;
     B.W. Harris, hep-ph/9608379;
     B.W. Harris and J. Smith, Nucl. Phys. B452 (1995) 109, Phys. Lett. B353 (1995) 535
175. G. Ingelman, J. Rathsman and G.A. Schuler, Comp. Phys. Comm. 101 (1997) 135
176. CDHS Collab., H. Abramowic et al., Z. Phys. C15 (1982) 19;
     E531 Collab., N. Ushida et al., Phys. Lett. B206 (1988) 380
177. EMC Collab., J.J. Aubert et al., Nucl. Phys. B213 (1983) 31

178. H1 Collab., contrib. paper 275 to HEP97, Jerusalem 1997
179. B.W. Harris and J. Smith, Phys. Rev. D57 (1998) 2806
180. C. Peterson, D. Schlatter, I. Schmitt and P. M. Zerwas, Phys. Rev. D27 (1983) 105
181. G. Goldhaber et al., Phys. Rev. Lett. 3 (1959) 181;
     G. Goldhaber, S. Goldhaber, W. Lee and A. Pais, Phys. Rev. 120 (1960) 300
182. H1 Collab., C. Adloff et al., Z. Phys. C75 (1997) 437
183. G. Goldhaber, Proc. LESIP 1 Workshop, Bad Honnef, 1984, eds. D. K. Scott and R. M. Weiner, p. 115
184. J. D. Bjorken, Proc. XXIV Int. Symp. on Multiparticle Dynamics, Vietri-sul-Mare, Italy, eds. A. Giovannini, S. Lupia and R. Ugoccioni, p. 579
185. K. Kolehmainen and M. Gyulassy, Phys. Lett. B180 (1986) 203;
     M. G. Bowler, Particle World 2 (1991) 1
186. J. D. Bjorken, Phys. Rev. D27 (1983) 140;
     X. Artru and G. Menessier, Nucl. Phys. B70 (1974) 93;
     B. Andersson and W. Hofmann, Phys. Lett. B169 (1986) 364;
     M. G. Bowler, Z. Phys. C29 (1985) 617;
     X. Artru and M. G. Bowler, Z. Phys. C37 (1988) 293
187. B. Andersson and M. Rignér, LU TP 97-07, LU TP 97-28
188. A. Bialas, Nucl. Phys. A545 (1992) c; Act. Phys. Pol. B23 (1992) 561
189. DELPHI Collab., P. Abreu et al., Z. Phys. C63 (1994) 17;
     EHS/NA22 Collab., I.V. Ajinenko et al., Z. Phys. C61 (1994) 567;
     UA1 Collab., N. Neumeister et al., Z. Phys. C60 (1993) 633
190. UA1 Collab., C. Albajar et al., Phys. Lett. B226 (1989) 410
191. E735 Collab., T. Alexopoulos et al., Phys. Rev. D48 (1993) 1931
192. H1 Collab., S. Aid et al., Nucl. Phys. B445 (1995) 3
193. H1 Collab., C. Adloff et al., Nucl. Phys. B504 (1997) 3
194. ZEUS Collab., M. Derrick et al., Z. Phys. C67 (1995) 93
195. ZEUS Collab., J. Breitweg et al., Phys. Lett. B414 (1997) 428
196. ZEUS Collab., contrib. paper N-662 to HEP97, Jerusalem 1997
197. D. Kant for the H1 and ZEUS Collab., 27th. Intern. Conf. on Multiparticle Dynamics, Frascati 1997
198. K.H. Streng, T.F. Walsh and P.M. Zerwas, Z. Phys. C2 (1979) 237
199. V. A. Khoze and W. Ochs, Int. J. Mod. Phys. A12 (1997) 2949
200. Yu. L. Dokshitzer, V.A. Khoze, A.H. Mueller and S.I. Troyan, Rev. Mod. Phys. 60 (1982) 373
201. TPC/Two-Gamma Collab., H. Aihara et al., Z. Phys. C44 (1989) 357
202. DELPHI Collab., P. Aarnio et al., Phys. Lett. B240 (1990) 271
203. OPAL Collab., M.Z. Akrawy et al., Phys. Lett. B247 (1990) 617
204. V.A. Khoze, S. Lupia and W. Ochs, Phys. Lett. B394 (1997) 179
205. A. I. Golokhvastov, Sov. J. of Nucl. Phys. 27 (1978) 430, ibid. 30 (1979) 128
206. D. Graudenz, CERN-TH/96-52, Habilitationsschrift, Univ. Hamburg 1996
207. TASSO Collab., W. Braunschweig et al., Z. Phys. C45 (1989) 11, ibid. C47 (1990) 187
208. Mark II Collab., A. Petersen et al., Phys. Rev. D37 (1988) 1
209. AMY Collab., Y.K. Li et al., Phys. Rev. D41 (1990) 2675
210. DELPHI Collab., P. Abreu et al., Z. Phys. C73 (1997) 229
211. D. Graudenz, PSI-PR/97-20, Proc. Workshop "New Trends in HERA Physics", Schloß Ringberg, Tegernsee 1997, eds. B.A. Kniehl, G. Kramer and A. Wagner, p. 146
212. A.D. Martin, R.G. Roberts and W.J. Stirling, Phys. Lett. B354 (1995) 155
213. J. Binnewies, B.A. Kniehl and G. Kramer, Z. Phys. C65 (1995) 471, Phys. Rev. D52 (1995) 4947

214. This explanation has been brought up by E. De Wolf
215. H1 Collab., C. Adloff et al., Phys. Lett. B406 (1997) 256
216. PLUTO Collab., C. Berger et al., Z. Phys. C12 (1982) 297
217. K. Rabbertz, Dissertation RWTH Aachen, in prep.
218. S. Bethke, PITHA 97/37, to appear in Proc. "QCD Euroconference 97", Mont-pellier 1997, ed. S. Narison, p.54
219. Yu. L. Dokshitzer and B. Webber, Phys. Lett. B404 (1997) 321
220. R. P. Feynman, Photon and Hadron Interactions, W. A. Benjamin, New York 1972
221. ZEUS Collab., M. Derrick et al., Phys. Lett. B306 (1993) 158
222. H1 Collab., I. Abt et al., Z. Phys. C61 (1994) 59
223. ZEUS Collab., contrib. paper N-649 to HEP97, Jerusalem 1997;
     ZEUS Collab., J. Breitweg et al., DESY 98-038
224. T. Carli for the H1 Collab., Proc. EPS-HEP97 Conference, Jerusalem 1997
225. H1 Collab., T. Ahmed et al., Phys. Lett. B346 (1995) 415
226. ZEUS Collab., M. Derrick et al., Phys. Lett. B363 (1995) 201
227. H1 Collab., S. Aid et al., Nucl. Phys. B449 (1995) 3
228. H1 Collab., C. Adloff et al., Phys. Lett. B415 (1997) 418
229. H1 Collab., contrib. paper 247 to HEP97, Jerusalem 1997
230. ZEUS Collab., M. Derrick et al., Z. Phys. C67 (1995) 81
231. D. Mikunas for the ZEUS Collab., to appear in Proc. Workshop DIS97, Chicago 1997, eds. J. Repond and D. Krakauer
232. JADE Collab., W. Bartel et al., Z. Phys. C33 (1986) 23
233. S. Catani, Yu.L. Dokshitzer and B.R. Webber, Phys. Lett. B285 (1992) 291
234. H1 Collab., contrib. paper 246 to HEP97, Jerusalem
235. M. Seymour, Z. Phys. C62 (1994) 127
236. H.U. Bengtsson and T. Sjöstrand, Comp. Phys. Comm. 46 (1987) 43;
     T. Sjöstrand, Comp. Phys. Comm. 82 (1994) 74
237. ZEUS Collab., J. Breitweg et al., DESY 97-191
238. OPAL Collab., R. Akers et al., Z. Phys. C63 (1994) 197
239. CDF Collab., F. Abe et al., Phys. Rev. Lett. 70 (1993) 713;
     D0 Collab., S. Abachi et al., Phys. Lett. B357 (1995) 500
240. M. Weber for the H1 Collab., to appear in Proc. Workshop DIS97, Chicago 1997, eds. J. Repond and D. Krakauer
241. D. Graudenz, Phys. Lett. B256 (1992) 519; Phys. Rev. D49 (1994) 3291; Comp. Phys. Comm. 92 (1995) 65
242. T. Brodkorb and J. G. Körner, Z. Phys. C54 (1992) 519;
     T. Brodkorb and E. Mirkes, Z. Phys. C66 (1995) 141
243. T. Trefzger for the ZEUS Collab., Proc. Workshop DIS96 on "Deep Inelastic Scattering and Related Phenomena", Rome 1996, eds. G. D'Agostini and A. Nigro, p. 434
244. G. Grindhammer for the H1 and ZEUS Collab., Proc. Workshop DIS95 on "Deep Inelastic Scattering and QCD", Paris 1995, eds. J.F. Laporte and Y. Sirois, p. 295
245. K. Flamm for the H1 Collab., Proc. Workshop DIS96 on "Deep Inelastic Scattering and Related Phenomena", Rome 1996, eds. G. D'Agostini and A. Nigro, p. 451
246. H1 Collab., C. Adloff et al., DESY 98-087
247. H1 Collab., C. Adloff et al., DESY 98-075
248. T. Hadig, C. Niedzballa, K. Rabbertz and K. Rosenbauer, Proc. Workshop on "Future Physics at HERA", Hamburg 1995–1996, eds. A. De Roeck, G. Ingelman and R. Klanner, vol. 1, p. 524
249. NMC Collab., D. Allasia et al., Phys. Lett. B258 (1991) 493

250. T. Carli, hep-ph/9709240, Proc. Workshop "New Trends in HERA Physics", Schloß Ringberg, Tegernsee 1997, eds. B.A. Kniehl, G. Kramer and A. Wagner, p. 129
251. H1 Collab., contrib. paper 244 to HEP97, Jerusalem 1997
252. G. Lobo and F. Zomer for the H1 Collab., to appear in Proc. Workshop DIS97, Chicago 1997, eds. J. Repond and D. Krakauer
253. H1 Collab., S. Aid et al., Phys. Lett. B354 (1995) 494
254. ZEUS Collab., M. Derrick et al., Phys. Lett. B345 (1995) 576
255. H.L. Lai et al., Phys. Rev. D51 (1995) 4763
256. M. Kuhlen, MPI-PhE/95-19 (1995), hep-ex/9508014, Proc. Workshop DIS95 on "Deep Inelastic Scattering and QCD", Paris 1995, eds. J.F. Laporte and Y. Sirois, p. 345
257. M. Klasen, G. Kramer and B. Pötter, Eur. Phys. J. C1 (1998) 261
258. H. Jung, hep-ph/9709425, to appear in Proc. Madrid Workshop on "Low x Physics", Miraflores de la Sierra 1997
259. G. Schuler and T. Sjöstrand, Z. Phys. C68 (1995) 607; Phys. Lett. B376 (1996) 193
260. H1 Collab., C. Adloff et al., DESY 98-076
261. M. Drees and R. Godbole, Phys. Rev. D50 (1994) 3124
262. J. Bartels and H. Lotter, Phys. Lett. B309 (1993) 400
263. J. Kwieciński, A. D. Martin, P. J. Sutton and K. Golec-Biernat, Phys. Rev. D50 (1994) 217; Phys. Lett. B335 (1994) 220
264. A.H. Mueller, Nucl. Phys. B (Proc. Suppl.) 18C (1990) 125; J. Phys. G17 (1991) 1443
265. J. Bartels, A. De Roeck and M. Loewe, Z. Phys. C54 (1992) 635; W.K. Tang, Phys. Lett. B278 (1992) 363
266. M. Kuhlen, Phys. Lett. B382 (1996) 441
267. A.J. Askew, D. Graudenz, J. Kwieciński, and A.D. Martin, Phys. Lett. B338 (1994) 92
268. J. Bartels, V. Del Duca and M. Wüsthoff, Z. Phys C76 (1997) 75
269. J. Kwieciński, C. A. M. Lewis and A. D. Martin, Phys. Rev. D54 (1996) 6664
270. J. Kwieciński, C. A. M. Lewis and A. D. Martin, Phys. Rev. D57 (1998) 496
271. J. Kwieciński, S. C. Lang and A. D. Martin, Phys. Rev. D55 (1997) 1273, ibid. D54 (1996) 1874
272. B. Webber, Proc. Workshop on "Physics at HERA", Hamburg 1991, eds. W. Buchmüller and G. Ingelman, vol. 1, p. 285; K. Golec-Biernat, L. Goerlich and J. Turnau, DTP/97/110
273. Carl R. Schmidt, Phys. Rev. Lett. 78 (1997) 4531
274. H1 Collab., S. Aid et al., Z. Phys. C70 (1996) 609
275. ZEUS Collab., contrib. paper EPS 0391 to Int. Europhys. Conf. on HEP, Brussels 1995
276. N. Pavel for the ZEUS Collab., Proc. Workshop DIS96 on "Deep Inelastic Scattering and Related Phenomena", Rome 1996, eds. G. D'Agostini and A. Nigro, p. 502
277. ZEUS Collab., M. Derrick et al., Phys. Lett. B338 (1994) 483
278. P. Lanius, Dissertation, Univ. Hamburg 1994, MPI-PhE/94-26
279. M. F. Hess, Dissertation, Univ. Hamburg 1996, MPI-PhE/96-16
280. T. Haas and P.B. Kaziewicz, ZEUS note 95-017; O. Deppe, T. Haas and N. Pavel, ZEUS note 93-106
281. M. Kuhlen for the H1 and ZEUS Collab., MPI-PhE/94-23, DESY 94-191, Proc. 6th Rencontre de Blois, "The Heart of the Matter", Blois 1994, eds. J.-F. Mathiot and J. Tran Thanh Van, p. 187
282. J. Bartels, H. Lotter and M. Vogt, Phys. Lett. B373 (1996) 215

283. M. Kuhlen for the H1 and ZEUS Collab., MPI-PhE/97-24, hep-ex/9709030, to appear in Proc. Madrid Workshop on "Low x Physics", Miraflores de la Sierra 1997
284. E. Panaro, Dissertation, Univ. Hamburg 1998
285. M. Kuhlen, MPI-PhE/96-23, DESY 96-234, hep-ex/9611008 Proc. XIXth Workshop on "High Energy Physics and Field Theory", Protvino 1996, eds. V. A. Petrov, A. P. Samokhin and R. N. Rogalyov, p. 57
286. M. Kuhlen, MPI-PhE/96-18, hep-ex/9610004, Proc. Workshop on "Future Physics at HERA", Hamburg 1995–1996, eds. A. De Roeck, G. Ingelman and R. Klanner, p. 606
287. A. Edin, G. Ingelman and J. Rathsman, Phys. Rev. D56 (1997) 7317
288. CDF Collab., F. Abe et al., Fermilab-PUB-97/024-E, Fermilab-PUB-97/024-E, Phys. Rev. Lett. 75 (1995) 4358;
D0 Collab., S. Abachi et al., Fermilab-CONF-96/249-E
289. M. Kuhlen, Proc. Workshop DIS96 on "Deep Inelastic Scattering and Related Phenomena", Rome 1996, eds. G. D'Agostini and A. Nigro, p. 495
290. R. Engel, Proc. XXIXth Rencontre de Moriond 1994, ed. J. Tran Thanh Van, p. 321
291. H1 Collab., contrib. paper 280 to HEP97, Jerusalem 1997
292. J. Kwieciński, S. C. Lang and A. D. Martin, DTP/97/56, hep-ph/9707240;
S. C. Lang, PhD thesis, Univ. Durham 1997
293. H1 Collab., contrib. paper pa03-049 to ICHEP'96, Warsaw 1996
294. H1 Collab., contrib. paper 255 to HEP97, Jerusalem 1997
295. H1 Collab., C. Adloff et al., DESY 98-143
296. J. Kwieciński, A.D. Martin and P.J. Sutton, Phys. Rev. D46 (1992) 921
297. J. Bartels et al., Phys. Lett. B384 (1996) 300
298. S. Wölfle for the ZEUS Collab., to appear in Proc. Workshop DIS97, Chicago 1997, eds. J. Repond and D. Krakauer
299. T. Brodkorb and E. Mirkes, Z. Phys. C66 (1995) 141
300. T. Haas and M. Riveline, Proc. Workshop on "Future Physics at HERA", Hamburg 1995–1996, eds. A. De Roeck, G. Ingelman and R. Klanner, vol. 1, p. 594
301. J. Bartels, A. De Roeck and M. Wüsthoff, Proc. Workshop on "Future Physics at HERA", Hamburg 1995–1996, eds. A. De Roeck, G. Ingelman and R. Klanner, vol. 1, p. 598
302. ZEUS Collab., contrib. paper N-659 to HEP97, Jerusalem 1997
303. ZEUS Collab., J. Breitweg et al., DESY 98-050
304. A. Goussiou for the D0 Collaboration, Proc. EPS-HEP97 Conference, Jerusalem 1997
305. V.T. Kim and G.B. Pivovarov, Phys. Rev. D57 (1998) 1341
306. G. 't Hooft, Phys. Rev. Lett. 37 (1976) 8; Phys. Rev. D14 (1976) 3432
307. A. Belavin, A. Polyakov, A. Schwarz and Yu. Tyupkin, Phys. Lett. B59 (1975) 85
308. A. Ringwald, Nucl. Phys. B330 (1990) 1;
O. Espinosa, Nucl. Phys. B343 (1990) 310
309. I.I. Balitskii and V.M. Braun, Phys. Lett. B314 (1993) 237
310. A. Ringwald and F. Schrempp, DESY 94-197, hep-ph/9411217, Proc. of Int. Sem. "Quarks 94", Vladimir, Russia, 1994, eds. D. Y. Grigoriev et al., p. 170
311. A. Ringwald and F. Schrempp, DESY 96-203, hep-ph/9610213, to appear in Proc. IXth Int. Sem. "Quarks 96", Yaroslavl, Russia, 1996
312. A.I. Vainshtein, V.I. Zakharov, V.A. Novikov and M.A. Shifman, Sov. Phys. Usp. 25(4) (1982)

313. S. Moch, A. Ringwald and F. Schrempp, Nucl. Phys. B507 (1997) 134; and in prep.
314. A. Ringwald and F. Schrempp, DESY 96-125, hep-ph/9607238, Proc. Workshop DIS96 on "Deep Inelastic Scattering and Related Phenomena", Rome 1996, eds. G. D'Agostini and A. Nigro, p. 481
315. A. Ringwald and F. Schrempp, DESY 97-115, hep-ph/9706399, to appear in Proc. Workshop DIS97, Chicago 1997, eds. J. Repond and D. Krakauer
316. S. Moch, A. Ringwald and F. Schrempp, DESY 97-114, hep-ph/9706400, to appear in Proc. Workshop DIS97, Chicago 1997, eds. J. Repond and D. Krakauer
317. T. Carli, M. Gibbs, M. Kuhlen, A. Ringwald and F. Schrempp, in preparation (version used: QCDINS 1.4.1)
318. T. Carli and M. Kuhlen, Nucl. Phys. B511 (1998) 85
319. M. Gibbs, T. Greenshaw, D. Milstead, A. Ringwald and F. Schrempp, Proc. Workshop on "Future Physics at HERA", Hamburg 1995–1996, eds. A. De Roeck, G. Ingelman and R. Klanner, vol. 1, p. 509
320. V. Kuvshinov and R. Shulyakovsky, Act. Phys. Polon. B28 (1997) 1629
321. G. Grindhammer, hep-ph/9709255, Proc. Workshop "New Trends in HERA Physics", Schloß Ringberg, Tegernsee 1997, eds. B.A. Kniehl, G. Kramer and A. Wagner, p. 37
322. N. I. Kochelev, hep-ph/9710540v2

Index

Springer Tracts in Modern Physics

* Denotes a Volume which contains a Classified Index starting from Volume 36

Springer Tracts in Modern Physics

Springer Tracts in Modern Physics

Springer
and the
environment

At Springer we firmly believe that an international science publisher has a special obligation to the environment, and our corporate policies consistently reflect this conviction.
We also expect our business partners – paper mills, printers, packaging manufacturers, etc. – to commit themselves to using materials and production processes that do not harm the environment. The paper in this book is made from low- or no-chlorine pulp and is acid free, in conformance with international standards for paper permanency.

 Springer